SPACOMM 2018

The Tenth International Conference on Advances in Satellite and Space Communications

April 22 - 26, 2018

Athens, Greece

SPACOMM 2018 Editors

Timothy Pham, Jet Propulsion Laboratory/ California Institute of Technology, USA
Krishna Pande, ICST and EE, National Chiao Tung University (NCTU), Taiwan
Peng Deng, The Pennsylvania State University (PSU), USA
Emmanouel Michailidis, University of Piraeus, Greece

Printed from e-media with permission by:

Curran Associates, Inc.
57 Morehouse Lane
Red Hook, NY 12571

Some format issues inherent in the e-media version may also appear in this print version.

Printed by Curran Associates, Inc. (2018)

International Academy, Research, and Industry Association (IARIA)
412 Derby Way
Wilmington, DE 19810

Phone: (408) 893-6407
Fax: (408) 527-6351

petre@iaria.org

Additional copies of this publication are available from:

Curran Associates, Inc.
57 Morehouse Lane
Red Hook, NY 12571 USA
Phone: 845-758-0400
Fax: 845-758-2633
Email: curran@proceedings.com
Web: www.proceedings.com

SPACOMM 2018

Forward

The Tenth International Conference on Advances in Satellite and Space Communications (SPACOMM 2018), held between April 22, 2018 and April 26, 2018 in Athens, Greece, continued a series of events to evaluate the state of the art in academia and industry on the satellite, radar, and antennas based communications bringing together scientists and practitioners with challenging issues, achievements, and lessons learnt.

Significant efforts have been allotted to design and deploy global navigation satellite communications systems, satellite navigation technologies, applications, and services experience still challenges related to signal processing, security, performance, and accuracy. Theories and practices on system-in-package RF design techniques, filters, passive circuits, microwaves, frequency handling, radars, antennas, and radio communications and radio waves propagation have been implemented. Services based on their use are now available, especially those for global positioning and navigation. For example, it is critical to identify the location of targets or the direction of arrival of any signal for civilians or on-purpose applications; smarts antennas and advanced active filters are playing a crucial role. Also progress has been made for transmission strategies; multiantenna systems can be used to increase the transmission speed without need for more bandwidth or power. Special techniques and strategies have been developed and implemented in electronic warfare target location systems.

The conference had the following tracks:

- Wireless Technology for 5G Networks
- Satellite and Space communications
- Satellite/space communications-based applications
- Antenna/Radar systems and signal processing
- Satellite and Free Space Optical Communications

We take here the opportunity to warmly thank all the members of the SPACOMM 2018 technical program committee, as well as all the reviewers. The creation of such a high quality conference program would not have been possible without their involvement. We also kindly thank all the authors who dedicated their time and effort to contribute to SPACOMM 2018. We truly believe that, thanks to all these efforts, the final conference program consisted of top quality contributions.

We also gratefully thank the members of the SPACOMM 2018 organizing committee for their help in handling the logistics and for their work that made this professional meeting a success.

We hope that SPACOMM 2018 was a successful international forum for the exchange of ideas and results between academia and industry and to promote further progress in the field of satellite and space communications. We also hope that Athens, Greece, provided a pleasant environment during the conference and everyone saved some time to enjoy the historic charm of the city.

SPACOMM 2018 Chairs

SPACOMM Steering Committee
Timothy T. Pham, Jet Propulsion Laboratory - California Institute of Technology, USA
Stelios Papaharalabos, National Centre for Scientific Research "Demokritos", Greece
Cathryn Peoples, Ulster University & The Open University, UK

SPACOMM Industry/Research Advisory Committee
Michael Sauer, Corning Cable Systems, USA
Vittorio Dainelli, Rheinmetall Italia S.p.A., Italy
Brian Niehoefer, TÜV Informationstechnik GmbH, Germany

SPACOMM 2018
Committee

SPACOMM Steering Committee

Timothy T. Pham, Jet Propulsion Laboratory - California Institute of Technology, USA
Stelios Papaharalabos, National Centre for Scientific Research "Demokritos", Greece
Cathryn Peoples, Ulster University & The Open University, UK

SPACOMM Industry/Research Advisory Committee

Michael Sauer, Corning Cable Systems, USA
Vittorio Dainelli, Rheinmetall Italia S.p.A., Italy
Brian Niehoefer, TÜV Informationstechnik GmbH, Germany

SPACOMM 2018 Technical Program Committee

Ashfaq Ahmed, COMSATS Institute of Information Technology - Wah campus, Pakistan
Mohamed Al-Mosawi, University of Portsmouth, UK
David N. Amanor, North Carolina A&T State University, USA
Michael Atighetchi, Raytheon BBN Technologies, USA
Arash Behboodi, RWTH Aachen University, Germany
Mark Bentum, University of Twente, Netherland
Maria Dolores Cano Baños, Universidad Politécnica de Cartagena. Spain
Genshe Chen, Intelligent Fusion Technology Inc., USA
Yao-Yi Chiang, University of Southern California, USA
Vittorio Dainelli, Rheinmetall Italia S.p.A., Italy
Peng Deng, The Pennsylvania State University, USA
Suleyman Eken, Kocaeli University, Turkey
Ibrahim Hokelek, TUBITAK BILGEM, Turkey
Suk-Seung Hwang, Chosun University, Korea
Adil Hakeem Khan, Nation College of Engineering and Tech., Guna MP, India
Emmanouel T. Michailidis, University of Piraeus, Greece
Kresimir Malaric, University of Zagreb, Croatia
Brian Niehoefer, TUV Informationstechnik GmbH, Germany
Nele Noels, University of Gent, Belgium
Krishna Pande, National Chiao Tung University, Taiwan
Stelios Papaharalabos, National Centre for Scientific Research "Demokritos", Athens, Greece
Cathryn Peoples, Ulster University & The Open University, UK
Timothy T. Pham, Jet Propulsion Laboratory - California Institute of Technology, USA
Ermanno Pietrosemoli, Abdus Salam International Centre for Theoretical Physics, Italy
Cong Pu, Marshall University, Huntington, USA
Vincent Roca, Inria research institute, France
Alexandru Rusu, University Politehnica of Bucharest, Romania
Michael Sauer, Corning Incorporated, USA
Cristian Stanciu, University Politehnica of Bucharest, Romania

Piotr Tyczka, ITTI Sp. z o.o., Poznań, Poland
Ouri Wolfson, University of Illinois, USA
Xiaozhou Yu, Northwestern Polytechnical University, China
Qiang Zhu, University of Michigan, USA

Copyright Information

Table of Contents

Multi-User Scheduling for Narrow Band Internet of Things

Shu-Feng Cheng*, Hsin-An Hou†, Li-Chun Wang‡, Kai-Ten Feng§ and Jen-Yuan Hsu¶
*‡§Department of Electrical and Computer Engineering
National Chiao Tung University, Hsinchu, Taiwan
†¶Industry and Technology Research Institute, Taiwan
*Email: cherishu999.eed02g@g2.nctu.edu.tw
†Email: hsinan_hou@itri.org.tw
‡Email: lichun@g2.nctu.edu.tw
§Email: ktfeng@mail.nctu.edu.tw
¶Email: jyhsu@itri.org.tw

***Abstract*—In this paper, we investigate the performance issue of massive Internet of Things (IoT) by evaluating to what extent the number of devices can be accommodated. We build a system level simulator based on the 3GPP TR 45.820 Narrow-Band IoT standard. We find that the current First In First Out (FIFO) approach can serve 630 devices per base station assuming that one device can be served at a time. Then, we develop the scheduling techniques - maximal delay tolerance and α-maximum tolerance for a system-level performance analysis. Our experiment results show that the proposed scheduling techniques outperform at least 2.8 times compared with the FIFO policy.**

Keywords–NB-IoT; Scheduling; Resource allocation.

I. Introduction

The Internet of Things (IoT) aims to enable all devices (i.e., electronics, transportation, sensors and home appliances) to communicate to each other without requiring human or computer interactions. By analyzing and exchanging the collected data from these IoT devices, various services can be integrated and be leveraged with each other to improve user experiences. The Fifth Generation (5G) communications systems provide faster speed and higher capacity than current 4G systems. Therefore, the IoT market is expected to have 30 billion connected devices by 2025 [1].

Low-Power Wide-Area Network (LPWAN) is one of the wireless communications for massive IoT devices. Some technologies, such as Long Range Wide Area Network (LoRaWAN), Sigfox and NB-IoT are designed to support a long range service for the devices subject to the constraint of battery life. However, LoRaWAN and Sigfox are proprietary product. Any non-mobile operator customers can utilize them to deploy the private network on the unlicensed frequency band. Therefore, these two technologies are suitable for the uncentralized local deployment. To completely realize the scenario of everything connected, we need the centralized signaling control to manage the data collected by the sensor. Therefore, the Narrowband IoT (NB-IoT), which is a cellular IoT, is proposed by the 3rd Generation Partnership Project (3GPP) [2]. It follows the LTE standards to design some part of the new specification for NB-IoT. In this way, the NB-IoT can be integrated in the existing LTE specifications [3] to coexist with the LTE system. The technical details tightly related to the LTE specifications are presented by Rohde & Schwarz [4]. In [5], the authors have evaluated and analyzed that the NB-IoT system can provide the extended coverage of 20 dB. The device and network requirements of the NB-IoT system are listed as follows:

- Extreme low data rate support $\sim$ few kbps
- Long delay tolerance $<$ 10s
- Ultra-low complexity
- Long Battery Life $>$ 10 years
- Low Cost $<$ 5 USD
- High Cell Capacity $\sim$ 52k user/cell site
- Extended Coverage link budget $\sim$ 164 dB Maximum Coupling Loss (MCL)
- Low Deployment and Operation Cost
- Consistent and Meaningful User Experience

The authors in [6] have proposed a link-level performance analysis with their inner loop and outer loop algorithms. However, the system-level performance have not been developed to evaluate the NB-IoT system before we deploy it. Additionally, the cell capacity for NB-IoT is still an open issue. Because the feature of repetitions for data and control signaling in NB-IoT, repeating data transmission will decrease the system capacity and even reduce the number of served UEs.

However, the frame structure in the uplink is different from the legacy LTE. A new Resource Unit (RU) is designed for NB-IoT. Besides, the scheduling delay is required in the data transmission because of the ultra-low complexity for NB-IoT devices. It does not consider for scheduling in the LTE system. Therefore, we build a simulation platform to estimate the total number of scheduled NB-IoT UEs with considerations of both repetition and multiple users subject to the latency requirement.

We propose the scheduling algorithms to observe the impacts of performances metrics (e.g., UE number and system throughput) among different scheduling policy. The scheduling algorithms are briefly summarized in the following.

- Maximum delay tolerance: UE with early-transmitted time is served first based on the maximum delay tolerance.
- α-maximum tolerance: Only consider the data transmission time of UEs is lower than the half of maximum delay tolerance. Policy is same as Maximum delay-tolerance.

The remainder of this paper are organized as follows. Section II presents the two-dimension multi-user scheduling and the proposed scheduling algorithms. In Section III, we introduce the flow chart of the simulator and numerical results are also discussed. Finally, we give our concluding remarks in Section IV.

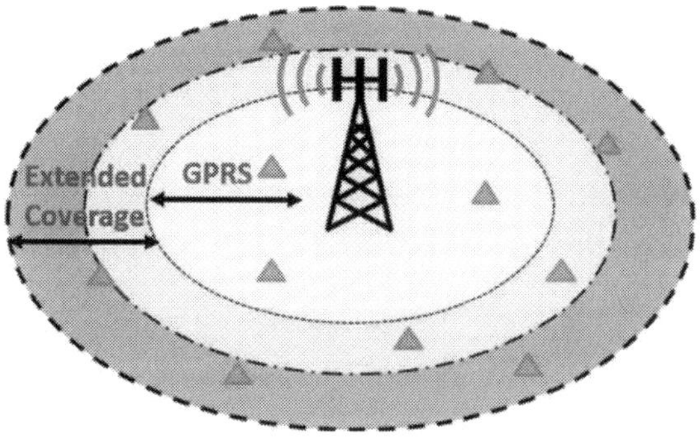

Figure 1. Cell architecture of NB-IoT system.

II. Two-dim Multi-User Scheduling

Consider the uplink transmission of the NB-IoT system, as shown in Figure 1. N uniformly distributed User Equipment (UE) with one omni-direction antenna are served by a base station. The Maximum Coupling Loss (MCL) between the transmitter and the receiver is 164 dB for NB-IoT system, which is 20 dB more than General Packet Radio Service (GPRS).

The piece-wise linear model is adopted for numerical analysis. The path loss model has been defined as follows:

$$P(L)_{dB} = \begin{cases} -K + 10\alpha_1 \log_{10}(d/d_0) & , d_0 \le d \le d_c \\ -K + 10\alpha_1 \log_{10}(d_c/d_0) & \\ \quad + 10\alpha_2 \log_{10}(d/d_c) & , d > d_c \end{cases} \tag{1}$$

where $K = 20\log_{10}\frac{\lambda}{4\pi d_0}$ is a constant path loss factor, d_0 is a reference distance, $d_c = \frac{4\pi h_t h_r}{\lambda}$ is a critical distance, α_1 and α_2 are the path loss exponents depending on the distance from eNB to the specific UE. Note that h_t and h_r are the antenna height of the transmitter and that of the receiver, respectively.

A. SNR-MCS Index Mapping

According to the above propagation model, the Signal-Noise-Ratio (SNR) can be measured at the eNB. For higher transmission efficiency and lower error rate, the eNB indicates the UEs to adapt the Modulation and Coding Scheme (MCS) index based on the corresponding Channel Quality Indicator (CQI). However, there is no CQI reporting for NB-IoT. Thus, we construct the SNR-to-MCS index mapping in Table I.

B. Repetitions Decision

To enhance the coverage performance, a NB-IoT system repeats its transmitted data transmission to combat the signal power attenuation. When the repetition number is doubled, the data rate is halved and the coverage gain is increased by 3 dB. Therefore, we decide the repetition number for the different channel of each UE based on the bit error rate for BPSK or QPSK modulation scheme. The repetition times for the data transmission of UE k can be expressed as follows:

$$r_k = 2^{\log_{\sqrt{2}} \frac{Q^{-1}(P_b)}{\sqrt{2\gamma_k}}}, \tag{2}$$

where P_b is a desired target bit error rate, γ_k is the SNR of UE k and $Q^{-1}(\cdot)$ represents the inverse Q function. The maximum repetition times is 2048.

TABLE I. SNR AND MCS INDEX MAPPING

Range of SNR values		MCS index
Lower limit	Upper limit	Value
-5	-4.16	0
-4.16	-3.32	1
-3.32	-2.48	2
-2.48	-1.64	3
-1.64	-0.80	4
-0.80	0.04	5
0.04	0.88	6
0.88	1.72	7
1.72	2.56	8
2.56	3.40	9
3.40	4.24	10
4.24	5.08	11
5.08	5.92	12

TABLE II. MCS INDEX AND SUBFRAME INDEX MAPPING

MCS	I_{SF}							
	0	1	2	3	4	5	6	7
0	16	32	56	88	120	152	208	256
1	24	56	88	144	176	208	256	344
2	32	72	144	176	208	256	328	424
3	40	104	176	208	256	328	440	568
4	56	120	208	256	328	408	552	680
5	72	144	224	328	424	504	680	872
6	88	176	256	392	504	600	808	1000
7	104	224	328	472	584	712	1000	
8	120	256	392	536	680	808		
9	136	296	456	616	776	936		
10	144	328	504	680	872	1000		
11	176	376	584	776	1000			
12	208	440	680	1000				

C. Subframe Index Selection

The BS obtains the MCS index based on the SNR value. Then, the UL data transmission time will be determined by the MCS index and subframe index. The mapping table is listed in Table II defined in [7]. The values in the table indicate that how many packet sizes can be transmitted subject to the MCS and subframe index. The subframe index is selected for the minimum value that greater than transmitted packet sizes corresponding to the MCS index.

D. Scheduling Schemes

For evaluating the total scheduled UEs for NB-IoT system, some scheduling algorithms are employed to allocate the resource to the multiple UEs. We adopt the Minimum Transmission Time (MTT) and First-In First-Out (FIFO) as two basis of the scheduling schemes. The MTT policy means the UEs with the minimum transmission time has higher priority to be scheduled. The FIFO policy let the first-arriving UE will be able to get the best choices. Moreover, we propose two scheduling policies with the consideration of the latency

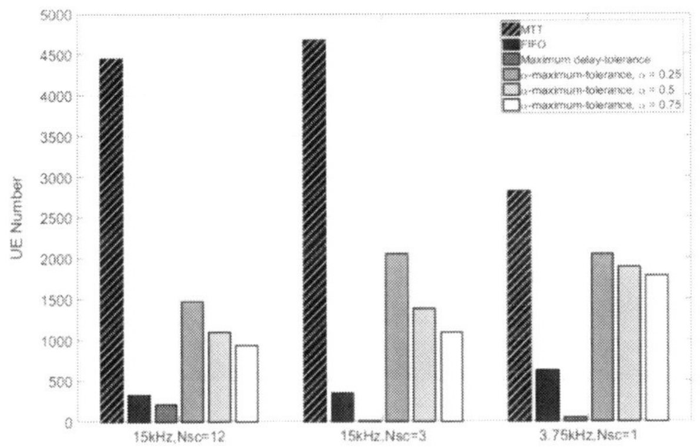

Figure 2. User number of NB-IoT system.

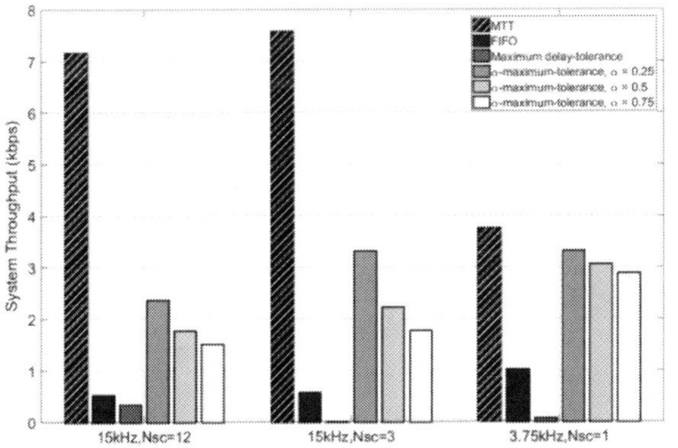

Figure 3. System throughput of NB-IoT system.

requirement. According to the delay requirement, we obtain the transmitted time point for each of UE in the buffer. As long as the UEs be scheduled before that time point, the delay requirement will be satisfied. Therefore, we propose the maximum delay-tolerance policy. The UE with early-transmitted time point is served first based on the maximum delay tolerance (i.e., 10 sec). However, the UEs with long data transmission time dominantly occupy the resource and decrease the system performances of overall throughput and served user number. To further enhance the system performances, we also propose the improved scheduling policy called α-maximum-tolerance. The parameter of α is determined by the ratio of the maximum delay tolerance. Only the data transmission time of UEs lower than the α of the maximum delay tolerance can be scheduled to utilize the resource. We adopt these scheduling policies to evaluate the system performance for the NB-IoT.

III. NUMERICAL RESULTS

Figure 6 shows the flow chart of our simulator. Table III summarizes the used system parameters [8]–[10]. The UEs are uniformly distributed in a single cell with the desired MCL of 144dB. Only one UE can be scheduled to use the system resource. We assume that all the resources are allocated to the UEs without considering the control signal overhead. The traffic modeladopted in our simulator is the Mobile Autonomous

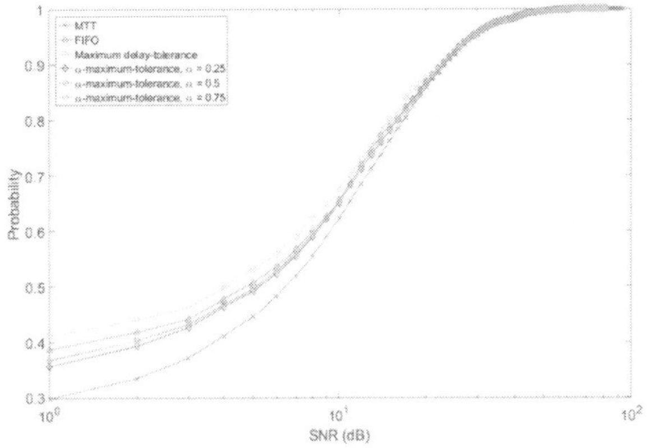

Figure 4. SNR CDF for different scheduling policy.

Reporting (MAR) periodic traffic. The packet sizes for the UEs are the random variable of Pareto distribution with the shape parameter $\beta = 1.5$. The minimum and the maximum packet sizes are 20 and 200 bytes, respectively. Additionally, the mapping table and the channel model are mentioned in Section II. Then, we discuss the simulation results of NB-IoT system by using the aforementioned simulation platform. Figure 2 shows the average total number of served UE for MTT, FIFO, maximum delay-tolerance and α-maximum-tolerance with $\alpha = 0.25, 0.5,$ and 0.75 scheduling policies. The maximum number of the served UEs for these policies are 4,746, 630, 208, 2,054, 1,894 and 1,788, respectively. For MTT policy, the short-transmission-time UE numbers are more than the long-transmission-time UE numbers as shown in Figure 5(a). Although the long-transmission-time UE is not many, the inner repetition for the UL transmission quality results in more resource utilization. Type-2 RU has less subcarriers than Type-1 RU. When Type-2 RU is adopted, the higher MCS level is selected. Besides, both the repetition and the inner repetition numbers will decrease , thereby freeing the resources to serve other UEs. However, once the band is narrow enough with the better channel condition, using Type-3 RU is unnecessary The frame structure of ype-3 RU is a long length of 8 ms time duration. Thus, the UE numbers decreases as the Type-3 RU is used for the MTT policy. On the contrary, the long-transmission-time UE has higher probability to be allocated the resources as shown in Figure 5(b) for the FIFO policy. In this way, the total UE number for FIFO is 10 times lower than that for MTT. Also, the advantage of Type-2 RU is canceled by them. When the Type-3 RU is used, the bandwidth is so narrow that it provides the higher power spectrum density (PSD) gain with same UL transmit power. Then, the total served UE number increases a little bit. Other policies are all delay awareness according to the different levels of the delay tolerance. Based on the differences of these levels, the proportion of the long-transmission-time UE are diverse as shown in Figure 5(c) and (d). Moreover, that results in variations of the total served UE numbers for distinct RU types. The average system throughput for the MTT, FIFO and proposed policies are shown in Figure 3. Because of the low packet size for NB-IoT devices, the system throughput is no more than 10 kbps. Figure 4 shows the CDF of SNR for the

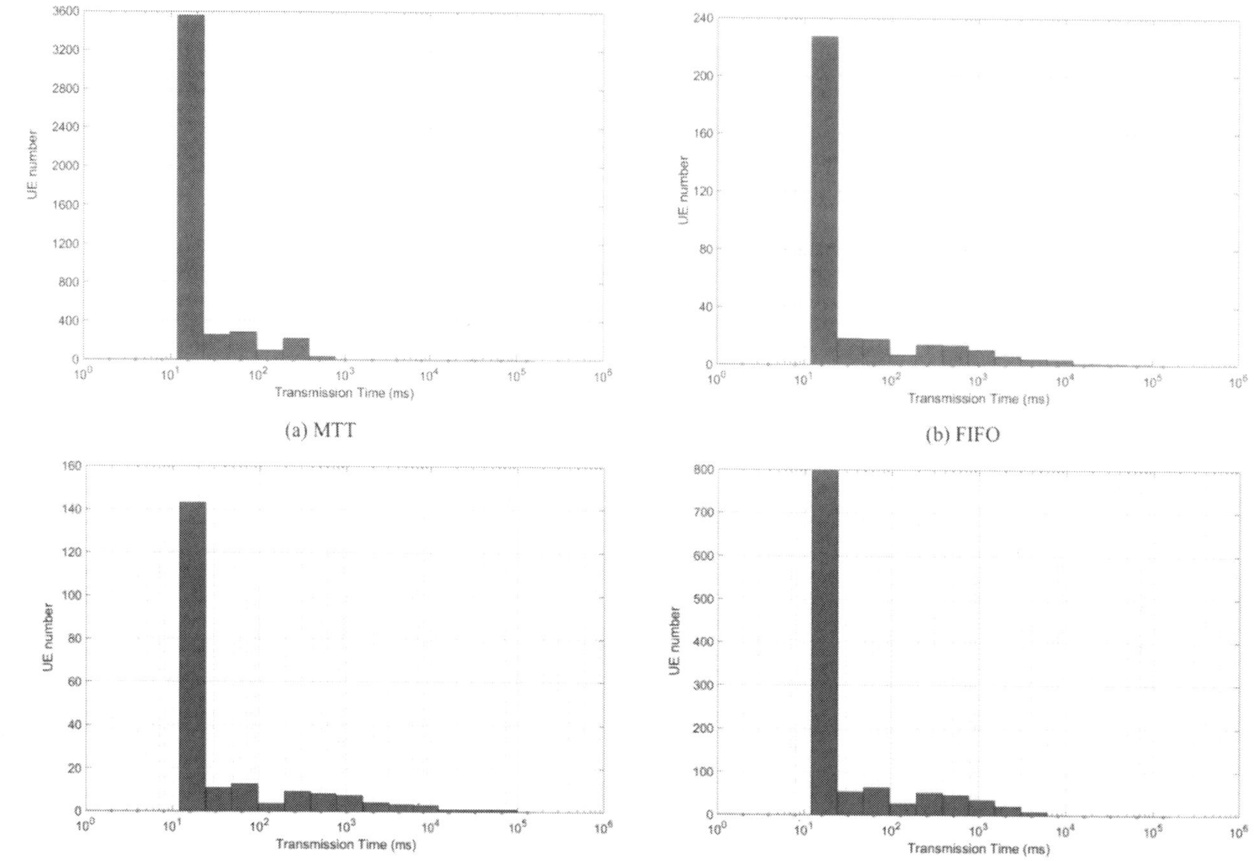

(a) MTT

(b) FIFO

(c) Maximum delay tolerance

(d) α-maximum-tolerance, $\alpha = 0.5$

Figure 5. Transmission time distribution.

TABLE III. SIMULATION PARAMETERS

Carrier Frequency	900 MHz
System Bandwidth	200 kHz
BS Transmit Power	24 dBm
UE Transmit Power	23 dBm
Noise Figure	3 dB
BS Height	25 m
UE Height	1.5 m
Subcarrier Spacing	{3.75, 15} kHz
Number of Sub-carriers	{1, 3, 6, 12} for 15 kHz SCS {1} for 3.75 kHz SCS
Shadowing Deivation	8 dB
Total Number of UEs	5000

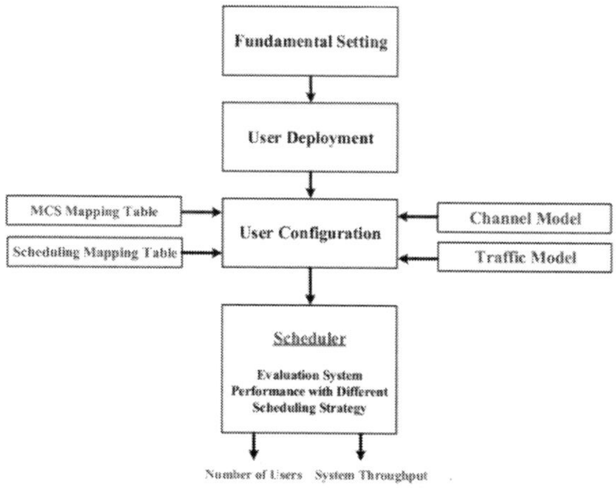

Figure 6. Flow chart of simulation platform.

served UE. We observe that the MTT policy is unfair for the UEs located at the worse environment, i.e., basement or the cell edge. These UEs have no opportunity to be served. On the other hand, UEs at the cell edge can also be served in the FIFO and proposed policies.

IV. CONCLUSION

In this paper, we have developed a simulator to evaluate the system-level performance for the NB-IoT system from the viewpoints of latency awareness. We consider the features of the repetition with the multiple user uplink scheduling. The MTT policy is the best method for the UEs located in the cell center. However, the NB-IoT devices deployed at the cell edge cannot be served. To overcome that problem, we propose the fairness and the delay-aware methodology. The performance of the total served UEs for the proposed maximum delay-tolerance policy is worse than the FIFO method due to the consideration of latency limitation. To improve that problem, we further propose the α-maximum-tolerance policy. With adjusting the parameter α, the delay tolerance level will be changed to satisfy the devices requirements. Our experimental results show that the total scheduled UEs for the improved policy have been raised to 2.8$\sim$3.26 times as compared with the FIFO. Our proposed method of the α-maximum-tolerance can provide a fairer way to schedule the UEs. Therefore, the system-level simulator with the proposed scheduling policies provide the preliminary methodology to evaluate the capacity performance for the NB-IoT system.

ACKNOWLEDGMENT

This research was supported by Industry and Technology Research Institute of Taiwan under the 5G NB-IoT System Validation and Enhancement Project.

REFERENCES

[1] M. Luthra, R. Atri, M. Sadeghian, S. Malik and P. Rekhi, "Long Term Evolution for IoT (Narrow Band LTE-Cellular IOT) A Short Note on Design, Technology and Applications," 2016.

[2] 3GPP TR 48.520, "Technical Specification Group GSM/EDGE Radio Access Network-Cellular System Support for Ultra-Low Complexity and Low Throughput Internet of Things (CIoT), V13.1.0 (Release 13)," Nov. 2015.

[3] Y. P. E. Wang, X. Lin, A. Adhikary, A. Grovlen, Y. Sui, Y. Blankenship, J. Bergman and H. S. Razaghi, "A Primer on 3GPP Narrowband Internet of Things (NB-IoT)," in IEEE Communications Magazine, Mar. 2017, pp. 117–123.

[4] J. Schlienz and D. Raddino, "Narrowband Internet of Things Whitepaper," 2016.

[5] A. Adhikary, X. Lin and Y. P. E. Wang, "Performance Evaluation of NB-IoT Coverage," 2016 IEEE 84th Vehicular Technology Conference (VTC-Fall), Sep. 2016, pp. 1–5.

[6] C. Yu, L. Yu, Y. Wu, Y. He and Q. Lu, "Uplink Scheduling and Link Adaptation for Narrowband Internet of Things Systems," in IEEE Access, Feb. 2017, pp. 1724–1734.

[7] 3GPP TS 36.213, "Evolved Universal Terrestrial Radio Access (E-UTRA); Physical layer procedures, V13.1.0 (Release 13)," Sep. 2016.

[8] 3GPP TS 36.211, "Evolved Universal Terrestrial Radio Access (E-UTRA); Physical channels and modulation, V13.1.0 (Release 13)," Sep. 2016.

[9] 3GPP TS 36.141, "Evolved Universal Terrestrial Radio Access (E-UTRA); Base Station (BS) conformance testing, V14.4.0 (Release 14)," June 2017.

[10] 3GPP TS 36.104, "Evolved Universal Terrestrial Radio Access (E-UTRA); Base Station (BS) radio transmission and reception, V14.4.0 (Release 14)," June 2017.

Realizing Challenging Internet of Things Applications Via Aerospace Infrastructures

Emmanouel T. Michailidis[1,2], Stelios M. Potirakis[2], and Athanasios G. Kanatas[1]
[1]Department of Digital Systems, University of Piraeus, Piraeus, Greece
e-mail: {emichail, kanatas}@unipi.gr
[2]Department of Electrical and Electronics Engineering, University of West Attica, Campus 2, Aigaleo, Athens, Greece
email: spoti@puas.gr

***Abstract*—This paper intends to justify why the aerospace communications is a key technology that will bring the Internet of Things (IoT) vision closer to reality. Both satellites and aerial platforms, such as High-Altitude Platforms (HAPs) and Unmanned Aerial Vehicles (UAVs), are considered. Their advantages are highlighted mainly in terms of the radio coverage and the link reliability, and their potential applications in several sectors are described. Emphasis is also given on the advanced wireless technologies that can further enhance the capabilities of aerospace communications.**

Keywords-High-Altitude Platforms (HAPs); Internet of Things (IoT); satellite networks; Unmanned Aerial Vehicles (UAVs).

I. INTRODUCTION

The Fifth-Generation (5G) vision does not only represent a significant upgrade of mobile broadband communications, but it will bring new unique network and service capabilities towards the evolution of Internet of Things (IoT) [1]. The IoT is an information network that encompasses a large family of smart and critical applications and supports a massive number of small, low-cost and low-power interconnected physical objects, generally referred to as Machine-Type Devices (MTDs), which interact and cooperate without human intervention, in order to reach common goals. More specifically, the IoT comprises sensors and actuators interacting with data processing, wired and short- and long-range wireless communication technologies, ground-based processing, and humans.

The successful operation of IoT-based networks envisages the seamless and synergetic integration of heterogeneous terrestrial and aerospace networks with different capabilities, which give rise to new services, architectures, and challenges [2]-[4]. From a communications engineering standpoint, the heterogeneity of the aforementioned technologies translates into highly different service requirements in terms of data traffic, end-to-end communication delay, reliability, etc. Cloud and fog computing and networking can facilitate the seamless integration of different heterogeneous networks, whereas Software-Defined Radios (SDR) and Software-Defined Networking (SDN) could bring flexibility and support the cost-efficient deployment and runtime of customized networks [5]. Motivated by these observations, this paper tries to shed light on the aerospace-based IoT and describe its special characteristics and the benefits provided by the aerospace infrastructures.

The rest of the paper is organized as follows. Section II describes the types of aerospace communication technologies. Section III underlines the advantages of satellites and aerial platforms, while Section IV presents the advanced wireless technologies that intend to enrich their capabilities. Section V outlines the potential applications of aerospace-based IoT. Finally, conclusions are drawn in Section VI.

II. TYPES OF AEROSPACE COMMUNICATIONS

Hybrid satellite-terrestrial networks are a typical example of cooperation between different architectures [2]. As the demand for comprehensive broadband and broadcast/multicast high-speed wireless communication services (e.g., voice, data, and multimedia services), global coverage, and ubiquitous access has grown, satellite networks can strongly support terrestrial backhaul networks and provide extensive as well as uninterrupted radio coverage to stationary, portable, and mobile receivers at frequencies ranging from 100 MHz to 100 GHz, as well as at optical frequencies, using different orbital architectures. Satellite are spaceborne vehicles capable of bringing communications to sparsely populated or underdeveloped areas, and still maintaining exclusive status in traditional maritime and aeronautical markets from the wide area perspective due to their unique coverage features.

In recent years, the use of mobile airborne services via aerial High-Altitude Platforms (HAPs), i.e., airships powered by solar cells or aircrafts powered by fuel engine propulsion, flying in the stratosphere, and Low-Altitude Platforms (LAPs) or Low- and Medium-Altitude Platforms (LMAPs) flying at various altitudes in the troposphere to supplement wireless terrestrial infrastructure has been also suggested [6]. The latter includes the Unmanned Aerial Vehicles (UAVs), i.e., Remotely Piloted Vehicles (RPVs), drones, robot planes, and pilotless aircrafts [7]. These platforms are capable of providing ubiquitous wireless access over large coverage areas at low cost, while they attain network flexibility and adaptability due to their rapid deployment and movement on demand. Project Loon initiated by Google has intended to leverage high-altitude balloons for broadband services in remote locations [8], whereas Facebook [9] has attempted to deploy solar-powered drones to provide Internet access to underdeveloped areas. Moreover, Microsoft has deployed balloons in the stratosphere loaded with sensors and cameras, which are connected to the Azure IoT platform and send the telemetry to the field gateway through the Constrained Application Protocol (CoAP). Figure 1 demonstrates a hybrid communication network consisting of terrestrial and aerospace infrastructures for IoT applications.

III. Advantages of Aerospace Communications

By leveraging the aerospace platforms, the IoT connectivity can be enhanced based on the additional communication choice. High-speed geostationary and/or non-geostationary satellites can provide efficient backhauling of aggregated IoT traffic from multiple sites and transfer data from Low-Power Wide Area Network (LPWAN) devices to applications on cloud platforms. They can also enable multicast/broadcast and trucking of video, IoT and other data across a large coverage area or a central site, with further terrestrial distribution to local cell sites. The aerospace communications have the potential to promote the Internet of Remote Things (IoRT) [10] in locations with terrestrial network constraints, i.e., remote, highly mobile, highly dispersed of wide geographical areas (rural, maritime, aero, railway, vehicular), and support massive Machine-to-Machine (M2M) communications. The aerospace infrastructure is not only capable of interconnecting remoted or dispersed smart objects, but also ensuring effectual management of data-intensive applications, redundant connections at critical sites, reliability, low latency, and enhanced capacity. A typical example of IoRT is the Global Sensor Network (GSN) for remote environment observation, where massively connected IoT sensor networks are connected via Low Earth Orbit (LEO) satellites.

Satellites have a major role in assisting 5G networks to meet sub-1ms latency requirements by delivering commonly accessed content to mobile base stations and multi-casting content to caches located at individual cells, even in places without fiber. Although Geostationary Earth Orbit (GEO) satellites latency of 250 ms (500 ms round-trip) is acceptable for many 5G and IoT applications and is comparable with the RTT of a long terrestrial link (100-200 ms), Medium Earth Orbit (MEO) and LEO satellite constellations allow for meeting more stringent latency requirements in case of voice and video transmission.

The aerospace platforms are not only an important supplement for big data acquisition methods for IoT, but also can improve the network performance. Since the links of terrestrial systems are often blocked, the aerial platforms, especially HAPs and drones, have great potential to attain a higher chance of Line-of-Sight (LoS) communication with the ground users and thus enhance the coverage and connectivity. The UAVs can also easily move, have a flexible deployment, and can provide rapid, on-demand communications. The realization of IoRT and M2M communications can be further reinforced due to the investments in new ground segment technologies, such as small, electronically steerable, electronically steerable, and/or phased-array satellite transceivers, as well as the cost-effective CubeSat platforms [11] based on micro, nano- and pico- satellites, operating as access points.

IV. Advanced Wireless Technologies

Although the local connectivity in IoT is provided by means of a short-range radio access technology, i.e., Bluetooth Smart, ZigBee, IPv6 over Low-Power Wireless Personal Area Networks (6LoWPAN), and Z-Wave, long-range wireless connectivity can be achieved using conventional, i.e., cellular and satellite networks, or LPWAN technologies, e.g., Longe Range (LoRa), Sigfox, Ingenu, and Narrowband-IoT (NB-IoT) [12]. In the context of satellite-based IoT applications, Inmarsat was the first to join the LoRa Alliance as its first satellite member and combine a LoRa Wide Area Network (LoRaWAN) on the ground with a satellite mesh in the sky. Inmarsat and other service providers predominantly use the lower data rate L-band (1-2 GHz). While L-band systems will continue to support extremely low data-rate IoT applications, higher frequencies, e.g., the Ka- and Ku-bands, are positioned for future applications, as network traffic increases. With Ku-band, service providers can leverage the large number of open Ku-band satellites, which traditionally offer lower spectrum cost than L-band. Although using Ka- and Ku-bands may be beneficial, rain effects severely affect signal propagation in mm-wave frequencies and a strong, dominant LoS signal is required for sufficient coverage due to the severe attenuation of the Non-Line-of-Sight (NLoS) links.

The capabilities of aerospace networks in terms of the latency are in part due to the use of Adaptive Coding and Modulation (ACM) techniques as foreseen in the Digital Video Broadcasting – Satellite – Second Generation (DVB-S2) standard, which retains outbound throughput even during heavy weather conditions. To further improve the performance of aerospace systems, algorithms based on cognitive radio principles can be applied that enable dynamic spectrum access and agile waveform adaptation (from carrier aggregation to unlicensed spectrum sharing).

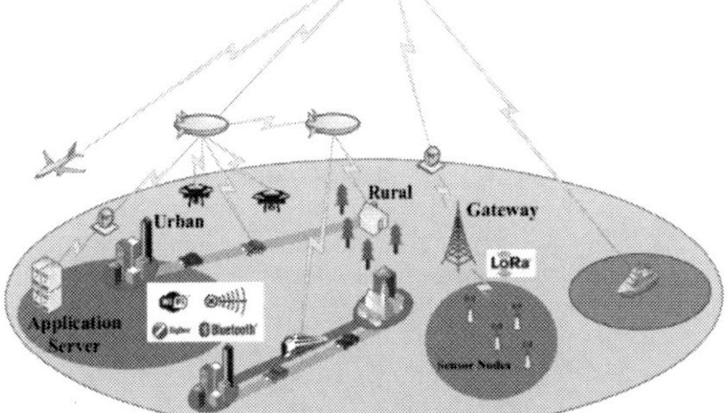

Figure 1. A simple representation of an hybrid terrestrial-aerospace communication network for IoT applications.

To successfully fulfill the growing demands for high data throughputs and enhanced link reliability with greater mobility support and extended network range, cooperative diversity preserves the end-to-end communication between a source and a destination via intermediate relays [13]. Since new requirements for enhanced spectral efficiency, greater bandwidth and higher data rates emerge, the Multiple-Input Multiple-Output (MIMO) [14] [15] and massive MIMO [16] technology can revolutionize satellite networks, exploit spatial diversity, and serve multiple users (i.e., multi-user MIMO). This has gained great interest due to the standardization activities on the finalized DVB - Satellite to Handheld (DVB-SH) standard [17], and the prospective DVB – Next Generation Handheld (DVB-NGH) [18] standard. As the performance of radar systems is limited by target scintillations, the application of MIMO techniques to Synthetic Aperture Radars (SARs) can greatly improves resolution and sensitivity, as well as detection and estimation performance of aerospace applications for the IoT by exploiting the diversity of target scattering [19].

Data rates in the multigigabit regime can be achieved using Free-Space Optical (FSO) inter-satellite, inter-platform, satellite-to-ground, and platform-to-ground connections [20]. Compared to Radio-Frequency (RF) systems, FSO systems are capable of providing intrinsic narrow beamwidth and high-speed LoS connectivity using the wavelength division multiplexing scheme with very small and compact equipment and low power consumption. In addition, FSO systems ensure privacy with low probability of interception, immunity to electromagnetic interference, and exemption from spectrum regulatory restrictions.

V. Potential Applications

It is important to note that the notion of IoT comprises a wide variety of applications. The challenging ones that will be benefited by the aerospace infrastructures refer to the critical IoT and require very high reliability and availability in terms of wireless connectivity. Although the average volume of data transported to and from devices may not be large, wide instantaneous bandwidths are useful in being able to meet capacity and latency requirements. On the other hand, the low cost of the devices and the energy consumption is not as important as for other IoT applications. Typical paradigms of critical IoT applications are the following:

- **Transport Applications:** The aerospace communications can improve the safety and management of transportation infrastructure and realize the Internet of Vehicles (IoV) [21]. Specifically, an expanded adoption of networking vehicles, sensors and controls can be achieved that will enable the smarter use of road and rail transportation by effectively supporting applications such as, signage, signaling and routing, alerts for road and weather conditions, level crossing protection, and train control systems. Satellite-based Global Positioning System (GPS) can determine the positioning of each vehicle, whereas HAPs represent a feasible solution for data acquisition in IoV, especially in rural areas. Besides, monitoring of vehicular traffics through drones with embedded cameras or sensors is also viable. By taking advantage of the aerospace infrastructure, the operation of commercial autonomous shipping can be enabled along with future commercial marine vessels, cargo logistics, and smart ports [22].
- **Industrial Applications:** The aerospace networks can also play a key role in a sub-segment of the IoT, referred to as the Industrial IoT (IIoT) [23], since they can deliver connectivity to remote locations and challenging environments. Although satellite networks are extensively deployed to support critical Supervisory Control And Data Acquisition (SCADA) applications, IIoT can be seen as an expansion of SCADA networks and other global asset tracking applications, remote control, automated fabrication, collaborative robots, etc. Applications tend to be those involving remote locations or extending over large land or sea areas, such as water level monitoring, oil and gas pipeline integrity, and tracking of mining trucks. Overall, the IIoT market seems a significant opportunity for satellite service providers.
- **Medical Applications:** The satellites and the aerial platforms combined with information and cloud technology are playing an increasing role in the support of welfare and Healthcare IIoT (HealthIIoT) applications [24] and can be used for providing low-cost and timely healthcare in remote and inaccessible areas. These technologies have a profound effect on the quality, safety, and efficiency of healthcare. The patients and the aging society can be served by satellite based telemonitoring, medical diagnosis and care from their homes, whereas interaction though videoconferencing between medical personnel and patients at home can be obtained. Moreover, using biomedical sensors, collection and transmission of medical data is feasible. Aerospace communications can also be used for monitoring endemics/epidemics at any area, while high bandwidth links can accommodate real-time medical imaging and remote robotic surgery [25].
- **Disaster and Crisis Management Applications:** As the natural disaster or large-scale unexpected events easily make the terrestrial network overloaded or totally destroyed, hybrid satellite-aerial-terrestrial networks have the ability to provide more effective services compared to traditional infrastructures during the emergency situations. Aerospace networks can offer rapid deployment and wide coverage and successfully support applications, such as video surveillance, structural monitoring and protection of critical infrastructure [26]. These include dams, bridges and other important structures,

as well as security and access controls related to border control, flood warning, earthquake detection, weather and environmental monitoring, coastline and pipeline surveillance, trafficability of maritime routes, e.g., icebergs, and synthetic aperture radar applications. The emergence of LAPs indicates a stable and reliable direction for the development of emergency network via circular flight tracks and continuous observation of regional hotspots and hazard areas.

- **Smart Energy Applications:** The aerospace infrastructure can efficiently handle a variety of electric grid automation and metering applications and support the integration and efficient use of all the energy resources and the grid infrastructure, using smart sensors, smart meters and smart control [27]. Moreover, there exists a wide range of applications in oil and gas that the aerospace infrastructure can effectively enable. These include monitoring and transmitting sensor data concerning drilling control, wellhead production, pipeline monitoring, distribution logistics and asset security.

VI. Conclusion

In this paper, the role of aerospace communication technologies in the future IoT ecosystem has been described. Since ubiquitous connectivity and long-range radio coverage are required in many critical IoT applications, satellites and aerial platforms along with advanced 5G wireless technologies can drastically change the landscape of various industries and strongly support the evolution of IoT.

Acknowledgment

This research is implemented through State Scholarships Foundation (IKY) and co-financed by the European Union (European Social Fund – ESF) and Greek national funds through the action entitled "Reinforcement of Postdoctoral Researchers" in the framework of the Operational Programme "Human Resources Development Program, Education and Lifelong Learning", with priority axis 6, 8, 9, of the National Strategic Reference Framework (NSRF) 2014-2020.

References

[1] M. R. Palattella et al., "Internet of Things in the 5G Era: Enablers, Architecture, and Business Models," *IEEE J. on Sel. Areas in Commun.*, vol. 34, no. 3, pp. 510-527, Mar. 2016.

[2] B. Evans et al., "Integration of satellite and terrestrial systems in future multimedia communications," *IEEE Wireless Commun.*, vol. 12, no. 5, pp. 72–80, Oct. 2005.

[3] B. Paillassa, B. Escrig, R. Dhaou, M.-L. Boucheret, and C. Bes, "Improving satellite services with cooperative communications," *Int. J. Satellite Commun. Netw.*, vol. 29, no. 6, pp. 479–500, Nov./Dec. 2011.

[4] B. G. Evans, "The role of satellites in 5G," in *Proc. 7th Advanced Satellite Multimedia Systems Conference and the 13th Signal Processing for Space Communications Workshop (ASMS/SPSC)*, Livorno, Italy, pp. 197-202, 2014.

[5] N. Zhang et al., "Software Defined Space-Air-Ground Integrated Vehicular Networks: Challenges and Solutions," *IEEE Commun. Magaz.*, vol. 55, no. 7, pp. 101-109, 2017.

[6] A. Mohammed, A. Mehmood, F. Pavlidou, and M. Mohorcic, "The role of high-altitude platforms (HAPs) in the global wireless connectivity," *Proc. IEEE*, vol. 99, no. 11, pp. 1939-1953, Nov. 2011.

[7] J. M. Sullivan, "Evolution or revolution? the rise of UAVs," *IEEE Technology and Society Magazine*, vol. 25, no. 3, pp. 43-49, Fall 2006.

[8] B. Moision et al. "Demonstration of free-space optical communication for long-range data links between balloons on Project Loon," in *Proc. SPIE 10096, Free-Space Laser Communication and Atmospheric Propagation XXIX*, 100960Z, Feb. 2017.

[9] Flying Aquila: Early lessons from the first full-scale test flight and the path ahead, 2016. [Online]. Available from: https://code.facebook.com/posts/268598690180189 [retrieved: March, 2018].

[10] M. De Sanctis, E. Cianca, G. Araniti, I. Bisio, and R. Prasad, "Satellite Communications Supporting Internet of Remote Things," *IEEE Internet of Things Journal*, vol. 3, no. 1, pp. 113-123, Feb. 2016.

[11] V. Almonacid and L. Franck, "Extending the coverage of the internet of things with low-cost nanosatellite networks", *Acta Astronautica*, Vol. 138, pp. 95-101, 2017.

[12] D. G. Kogias, E. T. Michailidis, G. Tuna, and V. C. Gungor, *Realizing the Wireless Technology in Internet of Things (IoT)*, in Emerging Wireless Communication & Network Technologies: Principle, Paradigm and Performance, Springer, in press.

[13] J. Jiao, H. Gao, S. Wu, and Q. Zhang, "Performance Analysis of Space Information Networks with Backbone Satellite Relaying for Vehicular Networks," *Wireless Communications and Mobile Computing*, vol. 2017, Article ID 4859835, 13 pages, 2017. doi:10.1155/2017/4859835

[14] P.-D. Arapoglou et al., "MIMO over Satellite: A Review," *IEEE Communications Surveys & Tutorials*, vol. 13, no. 1, pp. 27-51, First Quarter 2011.

[15] P.-D. Arapoglou, E. T. Michailidis, A. D. Panagopoulos, A. G. Kanatas, and R. Prieto-Cerdeira, "The Land Mobile Earth-Space Channel: SISO to MIMO Modeling from L- to Ka-Bands," *IEEE Vehicular Technology Magazine*, 6, 2, June 2011, pp. 44-53.

[16] E. G. Larsson, O. Edfors, F. Tufvesson, and T. L. Marzetta, "Massive MIMO for next generation wireless systems," *IEEE Communications Magazine*, vol. 52, no. 2, pp. 186-195, Feb. 2014.

[17] A. B. Alamanac et al., "In-Depth Analysis of the Satellite Component of DVB-SH: Scenarios, System Dimensioning, Simulations and Field Trial Results," *International Journal of Satellite Communications and Networking*, 27, 4-5, Jul.-Oct.2009, pp. 215-240.

[18] M. Sangchul et al., "Enhanced spatial multi plexing for rate-2 MIMO of DVB-NGH system," in *Proc. 19th International Conference on Telecommunications (ICT) 2012*, Jounieh, Lebanon, 23-25 April 2012, pp. 1-5.

[19] W. Wang, "Applications of MIMO Technique for Aerospace Remote Sensing," in *Proc. IEEE Aerospace Conference 2007*, Big Sky, MT, pp. 1-10, 2007.

[20] H. Kaushal and G. Kaddoum, "Optical Communication in Space: Challenges and Mitigation Techniques," *IEEE Communications Surveys & Tutorials*, vol. 19, no. 1, pp. 57-96, Firstquarter 2017.

[21] W. Xu et al., "Internet of vehicles in big data era," IEEE/CAA *Journal of Automat. Sinica*, vol. 5, no. 1, pp. 19-35, Jan. 2018.

[22] M. Höyhtyä, T. Ojanperä, J. Mäkelä, S. Ruponen, and P. Järvensivu, "Integrated 5G Satellite-Terrestrial Systems: Use Cases for Road Safety and Autonomous Ships," in *Proc. 23rd Ka and Broadband Communications Conference 2017*, Trieste, Italy, Oct. 2017.

[23] L. D. Xu, W. He, and S. Li, "Internet of Things in Industries: A Survey," *IEEE Transactions on Industrial Informatics*, vol. 10, no. 4, pp. 2233-2243, Nov. 2014.

[24] M. S. Hossain and G. Muhammad, "Cloud-assisted Industrial Internet of Things (IIoT) – Enabled framework for health monitoring," *Computer Networks*, vol. 101, pp. 192-202, Jun. 2016.

[25] S. Srivastava, M. Pant, A. Abraham, and N. Agrawal, "The Technological Growth in eHealth Services," Computational and Mathematical Methods in Medicine, vol. 2015, Article ID 894171, 2015. doi:10.1155/2015/894171

[26] Y. Wang, Y. Xu, Y. Zhang, and P. Zhang, "Hybrid satellite-aerial-terrestrial networks in emergency scenarios: a survey," *China Communications*, vol. 14, no. 7, pp. 1-13, July 2017.

[27] K. Sohraby, D. Minoli, B. Occhiogrosso, and W. Wang, "A Review of Wireless and Satellite-Based M2M/IoT Services in Support of Smart Grids," *Mobile Netw Appl*, 2017. https://doi.org/10.1007/s11036-017-0955-1

Optimal Relay Location and Opportunistic User-Scheduling for Stratospheric Communications

Emmanouel T. Michailidis[1,2], Nikolaos Nomikos[3], Petros Bithas[1,2], Demosthenes Vouyioukas[3], and Athanasios G. Kanatas[1]
[1]Department of Digital Systems, University of Piraeus, Piraeus, Greece
e-mail: {emichail, pbithas, kanatas}@unipi.gr
[2]Department of Electrical and Electronics Engineering, University of West Attica, Campus 2, Aigaleo, Athens, Greece
[3]Department of Information and Communication Systems Engineering, University of the Aegean, Karlovassi, Samos, Greece
email: {nnomikos, dvouyiou}@aegean.gr

Abstract—**This paper investigates the outage performance of a Decode-and-Forward (DF) Multi-User (MU) stratospheric relay communication system operating over Generalized-Gamma (GG) fading channels. This system consists of multiple users-sources, multiple users-destinations and a High-Altitude Platform (HAP) acting as a relay station in the stratosphere. Geometry-based optimal relay location with fixed power allocation is proposed, in order to minimize the outage probability. Exact and approximated expressions for the End-to-End (E2E) outage probability are derived and opportunistic scheduling is applied, where the Source-Destination (SD) pair with the highest instantaneous E2E Signal-to-Noise Ratio (SNR) is scheduled for communication. The results highlight the gains offered by the opportunistic SD scheduling and the optimization of the relay location on the overall performance.**

Keywords-Generalized Gamma (GG) fading; High-Altitude Platforms (HAPs); Multi-User (MU); oppurtunistic scheduling; outage probability; relay location.

I. INTRODUCTION

A key objective in the development of future generation communication systems and the Internet of Things (IoT) applications is the seamless integration of wireless terrestrial and aerospace infrastructures over heterogeneous networks [1] [2]. In recent years, a great interest has been drawn towards the development of High-Altitude Platforms (HAPs), intending to provide ubiquitous wireless access over large coverage areas at low cost, while attaining network flexibility and adaptability due to their rapid deployment and movement on demand [3]. The term HAPs defines aerial platforms flying in the stratosphere.

The radio links of terrestrial nodes are usually vulnerable to fading effects arising from scattering, reflection, diffraction, or blockage of the radiated energy by objects in the propagation environment. Hence, exploiting the features of cooperative diversity techniques by preserving the End-to-End (E2E) communication between a source (S) and a destination (D) via an intermediate relay (R) can fulfill the demands for enhanced link reliability with greater mobility support and extended network range [4].

With the rapid growth of the number of connected devices, Fifth-Generation (5G) communication systems will be deployed in dense-user environments. Hence, shifting from Single-User (SU) to Multi-User (MU) systems is indispensable [5]. In MU systems, the MU diversity gain can significantly improve the system performance by appropriately designing user scheduling schemes. These schemes are necessary, as a single relay may be incapable of simultaneously serving multiple users. The opportunistic scheduling is a well-known and efficient user selection schemes that are usually used to select among the users [6]. In this scheme, the user with the best channel conditions is always allowed to conduct its transmission in both downlink and uplink scenarios. Also, this scheme is usually used to achieve the maximum sum-rate capacity in wireless networks. However, in scenarios where asymmetric link conditions exist, a trade-off between capacity and fairness among users arises.

More recently, the investigation of relaying in Multi-Source (MS) Multi-Destination (MD) networks has been investigated [7], since based on these schemes substantial achievable rate improvement in shared-spectrum multiple access wireless systems are offered. In these networks, multiple Source-Destination (SD) pairs simultaneously communicate with the help of relay nodes. Typical paradigm of this configuration is the Interference Relay Channel (IRC) [8], where a relay assists two independent source-destination pairs by using different relaying methods, as well as the cellular operation, in which multiple users within a cell communicate with each other through a base station acting as a relay unit.

Motivated by the aforementioned observations, this paper investigates the use of a radio relay installed on a HAP that acts as a stratospheric relay station, between multiple user pairs. This configuration is envisioned to be applied in 5G systems, where the relay nodes are expected to be located at the cell edge establishing the communication links between cell-edge users. An opportunistic scheduling scheme is adopted, in order to select the pair with the best E2E channel among the available ones. In the proposed system, each user pair is able to communicate through the relay. Depending on the propagation environment, each link may experience different channel fading conditions. In particular, Rician, Rayleigh, and log-normal distributions are suggested to model the channel fading in urban, sub-urban, and rural environments, respectively [9]. By generalizing the previous approaches, this paper considers that the channel fading follows the Generalized-Gamma (GG) distribution [10], which includes many well-known distributions as special cases, whereas it has the important advantage of approximately describing other distributions [11]. Also, this paper proposes geometry-based optimization schemes regarding the position of the stratospheric relay for determining the SD pair with the best E2E Signal-to-Noise Ratio (SNR). The performance of the underlying system is analyzed in terms of the outage probability. The results demonstrate the effect of the number of SD pairs, fading characteristics, and optimized relay location on the system's performance.

The rest of the paper is organized as follows. Section II presents the system model and the geometrical characteristics, whereas Section III analyzes the channel statistics. In Section IV, mathematical expressions for the outage probability are derived. Section V proposes geometrical design recommendation for optimal outage performance. Results are provided in Section VI. Finally, conclusions are drawn in Section VII.

II. SYSTEM MODEL

This paper considers a MS-MD relay-assisted airborne communication system with slowly-varying, frequency-flat block-fading channels, where the channel coefficients stay constant over an entire block of communication. As shown in Figure 1, the proposed system consists of one cluster of K users, the sources, trying to communicate with one cluster of K users, the destinations, with the aid of a Decode-and-Forward (DF) stratospheric relay. This relay is situated approximately 20 km above the ground, equipped with two antennas, one for transmitting and the other for receiving. In this system, two users communicate with each other with the help of the relay since the direct links between them are obstructed due to high attenuation in the propagation medium. To aid our analysis, the subscripts S_k, R, and D_k, where $1 \le k \le K$, are affiliated with the k-th source, the relay, and the k-th destination, respectively. All of the user pairs are in the set U, where the k-th user pair is denoted by (S_k, D_k). Due to hardware limitations, the communication operates in a Half-Duplex (HD) mode and is conducted over two phases; the first phase (hop) includes the link between S_k and R, while the second phase (hop) involves the link between R and D_k. Time is divided in time-slots and multiple access is based on Time-Division Multiple Access (TDMA), where at each time-slot one SD pair is activated for transmission. The channel coefficients $h_{S_k,R}\left(h_{R,D_k}\right)$ between the k-th source (destination) and the relay are generalized gamma fading coefficients and identically distributed. In addition, in all cases the received signal is also affected by Additive White Gaussian Noise (AWGN) with variance N_O.

A. Geometrical Characteristics

For analytical simplification purposes, we assume that the users and the relay are placed on a x-y plane. Specifically, as shown in Figure 2, the users are placed on the horizontal axis, whereas the relay is placed on the vertical axis. This assumption is reasonable since it minimizes the effect of path loss. The fundamental parameters, which describe the geometry of the proposed system, are the elevation angle $\beta_{S_k}\left(\beta_{D_k}\right)$ of the relay relative to the k-th source (k-th destination), the height of the relay H_R, the distance $d_{S_k}\left(d_{D_k}\right)$ between the k-th source (k-th destination) and the relay, and the distance $D_{S_k}\left(D_{D_k}\right)$ from the k-th source (k-th destination) to the Sub-Platform Point (SPP). According to this geometry, the distance between the k-th source and the k-th destination is $D_{S_k D_k}$.

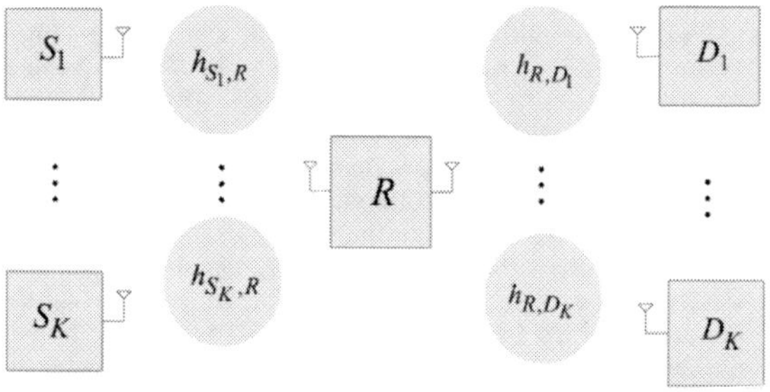

Figure 1. The MS-MD stratospheric relay fading channel.

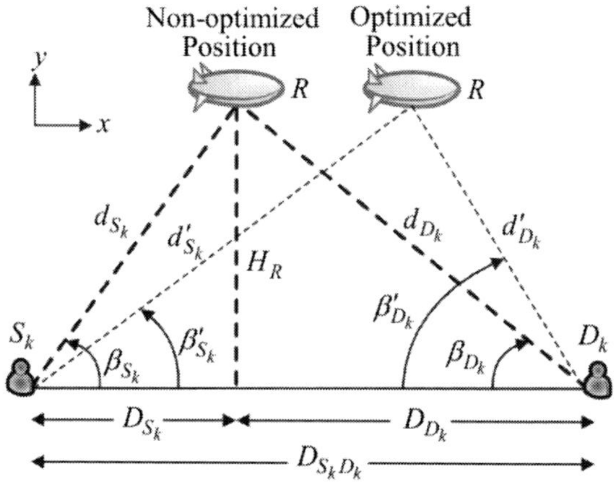

Figure 2. The geometrical characteristics of the proposed stratospheric relay communication system.

From Figure 2, one observes that d_{S_k} can be written as a function of $\beta_{S_k} \in (0, \pi/2)$ as follows

$$d_{S_k} = H_R / \sin \beta_{S_k}. \quad (1)$$

Using (1) and applying the cosine law to the triangle $S_k R D_k$ in Figure 2, d_{D_k} can be also written as a function of β_{S_k} as

$$d_{D_k} = \sqrt{\frac{H_R^2}{\sin^2 \beta_{S_k}} + D_{S_k D_k}^2 - 2\frac{H_R D_{S_k D_k}}{\tan \beta_{S_k}}}. \quad (2)$$

B. First Hop

The received signal at the stratospheric relay R from the kth source can be expressed as

$$y_{S_k,R} = \sqrt{P_{S_k}} h_{S_k,R} x_{S_k} + n_R, \quad (3)$$

where P_{S_k} is the transmit power at S_k, x_{S_k} is the transmitted symbol from the kth source with $\mathrm{E}\{|x_{S_k}|^2\} = 1$, $\mathrm{E}\{\cdot\}$ is the statistical expectation operator, and n_R represents the AWGN. The channel gain $|h_{S_k,R}|^2$ has mean power [12]:

$$\mathrm{E}\{|h_{S_k,R}|^2\} = d_{S_k}^{-a}, \quad (4)$$

where a is the path loss exponent. The instantaneous received SNR in the first hop is denoted as $\gamma_{S_k,R}$ and can be expressed as

$$\gamma_{S_k,R} = \frac{P_{S_k}}{N_O} |h_{S_k,R}|^2. \quad (5)$$

Using (4), the average SNR can be expressed as

$$\bar{\gamma}_{S_k,R} = \frac{P_{S_k}}{N_O} \mathrm{E}\{|h_{S_k,R}|^2\} = \frac{P_{S_k}}{N_O} d_{S_k}^{-a}. \quad (6)$$

C. Second Hop

The signal received at the destination can be written as

$$y_{R,D_k} = \sqrt{P_R} h_{R,D_k} y_{S_k,R} + n_{D_k}, \quad (7)$$

where P_R is the transmit power at R and n_{D_k} is the AWGN. The channel gain $|h_{R,D_k}|^2$ has mean power [12]:

$$\mathrm{E}\{|h_{R,D_k}|^2\} = d_{D_k}^{-a}. \quad (8)$$

The instantaneous received SNR in the second hop is denoted as γ_{R,D_k} and can be expressed as

$$\gamma_{R,D_k} = \frac{P_R}{N_O} |h_{R,D_k}|^2. \quad (9)$$

Using (8), the average SNR can be expressed as

$$\bar{\gamma}_{R,D_k} = \frac{P_R}{N_O} \mathrm{E}\{|h_{R,D_k}|^2\} = \frac{P_R}{N_O} d_{D_k}^{-a}. \quad (10)$$

Finally, the E2E SNR at the k-th destination can be approximated as follows

$$\gamma_{D_k} \approx \min\{\gamma_{S_k,R}, \gamma_{R,D_k}\}. \quad (11)$$

To analyze the performance of the proposed system, obtaining the statistics of the E2E SNR provided in (11) is required.

The scheduling is based on an opportunistic SD pair selection, where at each time-slot, the pair with the best E2E SNR is activated for communication. More specifically, the opportunistic SD pair selection will activate the k^*-th pair according to:

$$k^* = \arg \max_{1 \le k \le K} \min\{\gamma_{S_k,R}, \gamma_{R,D_k}\}. \quad (12)$$

In this way, the SD pair will be the one having the E2E channel offering the maximum capacity at each time-slot.

III. CHANNEL STATISTICS

In this paper, the GG distribution is utilized to model the channels for the S_k–R and R–D_k links. Hence, the Probability Density Function (PDF) of the instantaneous SNR received at R, denoted as $\gamma_{S_k,R}$, is given by [13]

$$f_{\gamma_{S_k,R}}(\gamma_{S_k,R}) = \frac{b_{S_k,R} \gamma_{S_k,R}^{m_{S_k,R} b_{S_k,R}/2 - 1}}{2\Gamma(m_{S_k,R}) (\tau_{S_k,R} \bar{\gamma}_{S_k,R})^{m b_{S_k,R}/2}} \times \exp\left[-\left(\frac{\gamma_{S_k,R}}{\tau_{S_k,R} \bar{\gamma}_{S_k,R}}\right)^{b_{S_k,R}/2}\right], \quad (13)$$

where $b_{S_k,R} > 0$ and $m_{S_k,R} \ge 1/2$ are the distribution's shaping parameters related to the fading severity, $\tau_{S_k,R} = \Gamma(m_{S_k,R}) / \Gamma(m_{S_k,R} + 2/b_{S_k,R})$, and $\Gamma(\cdot)$ is the Gamma function [14, eq. (8.310/1)]. For different values of $b_{S_k,R}$ and $m_{S_k,R}$, several distributions used in fading channel modeling can be obtained, e.g., Rayleigh (for $b_{S_k,R} = 2$ and $m_{S_k,R} = 1$), Nakagami-m (for $b_{S_k,R} = 2$), Weibull (for $m_{S_k,R} = 1$), and log-normal (for the limiting case of $b_{S_k,R} \to 0$ and $m_{S_k,R} \to \infty$). In addition, the PDF in (13) can approximately describe the Rician fading for $b_{S_k,R} = 2$ and $m_{S_k,R} = (K_{S_k,R} + 1)^2 / (2K_{S_k,R} + 1)$, where $K_{S_k,R}$ is the Rician factor of the first hop, i.e., the average power ratio of the Line-of-Sight (LoS) component to the Non-Line-of-Sight (NLoS) component.

The Cumulative Distribution Function (CDF) of the instantaneous SNR received at R can be expressed as

$$F_{\gamma_{S_k,R}}\left(\gamma_{S_k,R}\right)=1-\frac{\Gamma\left[m_{S_k,R},\left(\gamma_{S_k,R}/\left(\tau_{S_k,R}\bar{\gamma}_{S_k,R}\right)\right)^{b_{S_k,R}/2}\right]}{\Gamma\left(m_{S_k,R}\right)}, \quad (14)$$

where $\Gamma(\cdot,\cdot)$ stands for the upper incomplete Gamma function [14, eq. (8.350/2)]. Note that the PDF $f_{\gamma_{R,D_k}}\left(\gamma_{R,D_k}\right)$ and the CDF $F_{\gamma_{R,D_k}}\left(\gamma_{R,D_k}\right)$ of the GG distribution for the second RF link can be defined as in the first RF link by replacing the indices.

IV. Derivation of the Outage Probability

The outage probability is defined as the probability that the SNR at the destination goes below a predetermined outage threshold γ_{out}, i.e., $P_{out}=\Pr\left[\gamma_{D_k}\leq\gamma_{out}\right]$, where $\Pr[\cdot]$ is the probability operation. In this case, the communication system cannot achieve adequate reception. Using (11), the outage probability can be obtained from the CDF of the E2E SNR as $P_{out}=F_{\gamma_{D_k}}\left(\gamma_{out}\right)$ and can be written in terms of CDFs of the two hops' SNRs as follows [15]

$$P_{out}=1-\left[\left(1-F_{\gamma_{S_k,R}}\left(\gamma_{out}\right)\right)\left(1-F_{\gamma_{R,D_k}}\left(\gamma_{out}\right)\right)\right]. \quad (15)$$

From (15), one observes that the system gets in outage providing that at least one of the two hops gets in outage or, equivalently, the SNR of that hop becomes less than γ_{out}. Using (14), (15) becomes

$$\begin{aligned}P_{out}&=1-\frac{1}{\Gamma\left(m_{S_k,R}\right)\Gamma\left(m_{R,D_k}\right)}\\&\quad\times\Gamma\left[m_{S_k,R},\left(\gamma_{out}/\left(\tau_{S_k,R}\bar{\gamma}_{S_k,R}\right)\right)^{b_{S_k,R}/2}\right]\\&\quad\times\Gamma\left[m_{R,D_k},\left(\gamma_{out}/\left(\tau_{R,D_k}\bar{\gamma}_{R,D_k}\right)\right)^{b_{R,D_k}/2}\right]\\&=1-\left[\exp\left(-\left(\gamma_{out}/\left(\tau_{S_k,R}\bar{\gamma}_{S_k,R}\right)\right)^{b_{S_k,R}/2}\right)\right.\\&\quad\left.\times\sum_{k=0}^{m_{S_k,R}-1}\frac{1}{k!}\left(-\left(\gamma_{out}/\left(\tau_{S_k,R}\bar{\gamma}_{S_k,R}\right)\right)^{b_{S_k,R}/2}\right)^{k}\right]\\&\quad\times\left[\exp\left(-\left(\gamma_{out}/\left(\tau_{R,D_k}\bar{\gamma}_{R,D_k}\right)\right)^{b_{R,D_k}/2}\right)\right.\\&\quad\left.\times\sum_{k=0}^{m_{R,D_k}-1}\frac{1}{k!}\left(-\left(\gamma_{out}/\left(\tau_{R,D_k}\bar{\gamma}_{R,D_k}\right)\right)^{b_{R,D_k}/2}\right)^{k}\right], \quad (16)\end{aligned}$$

where $\Gamma(n,x)=\Gamma(n)\exp(-x)\sum_{k=0}^{n-1}\left(x^k/k!\right)$ for arbitrary integer n. At high SNRs and for the special case of Nakagami-m fading, using the approximation $\exp(-x)\approx 1-x$ for small x, (16) can be approximated as follows [12]

$$P_{out}\approx\frac{A}{\bar{\gamma}_{S_k,R}^{m_{S_k,R}}}+\frac{B}{\bar{\gamma}_{R,D_k}^{m_{R,D_k}}}, \quad (17)$$

where

$$A=m_{S_k,R}^{m_{S_k,R}}\gamma_{out}^{m_{S_k,R}}/\Gamma\left(m_{S_k,R}+1\right), \quad (18)$$

$$B=m_{R,D_k}^{m_{R,D_k}}\gamma_{out}^{m_{R,D_k}}/\Gamma\left(m_{R,D_k}+1\right). \quad (19)$$

Note that the aforementioned approximation is valid in a broad SNR regime [12].

Depending on the values of the parameters $b_{S_k,R}\left(b_{R,D_k}\right)$ and $m_{S_k,R}\left(m_{R,D_k}\right)$, the S_k–R and R–D_k links may experience either symmetric, e.g., Rayleigh/Rayleigh or Rician/Rician, or asymmetric, e.g., Rayleigh/Rician or Rician/Rayleigh, fading phenomena.

V. Optimal Location of the Stratospheric Relay

An important design parameter is to determine the optimal relay position that will minimize the outage probability for a given set of users. For this optimal position, the proposed scheduling approach determines the pairs of users that will communicate, according to the opportunistic scheduling principle.

In order to determine the optimal relay position, we set $x=\sin\beta_{S_k}$ and consider a predetermined power allocation $\left(P_{S_k}/N_O, P_R/N_O\right)$ and fixed values of H_R and D_{SD_k}. Thus, the problem of finding the optimal relay location with respect to S_k and D_k can be stated as follows

$$x^{*}=\arg\min_{x} P_{out}$$
$$\text{subject to: } 0<x<1 \quad (20)$$

Using (6), (10), and (17)-(19) and computing the second derivative of P_{out} with respect to x, one can easily show that the objective function is a strictly convex function of $x\in(0,1)$. Moreover, computing the first derivative of P_{out} with respect to x, the optimal relay location is the root of the equation in (21).

Standard iterative root-finding algorithms can be used to efficiently find numerical solutions to x^{*} and then compute $\beta_{S_k}=\arcsin x$. Then, using (2), we can compute d_{D_k} and $\beta_{D_k}=\arcsin\left(H_R/d_{D_k}\right)$.

$$
\begin{aligned}
&0.5aBm_{R,D_k}\left(-\frac{2H_R^2}{x^3}+\frac{2D_{S_kD_k}H_R}{\sqrt{1-x^2}}+\frac{2D_{S_kD_k}H_R\sqrt{1-x^2}}{x^2}\right)\\
&\times\left(D_{S_kD_k}^2+\frac{H_R^2}{x^2}-\frac{2D_{S_kD_k}H_R\sqrt{1-x^2}}{x}\right)^{-1+\frac{a}{2}}\left(\frac{P_R}{N_O}\right)^{-1}\\
&\times\left(\frac{\left(D_{S_kD_k}^2+\frac{H_R^2}{x^2}-\frac{2D_{S_kD_k}H_R\sqrt{1-x^2}}{x}\right)^{a/2}}{\left(\frac{P_R}{N_O}\right)}\right)^{-1+m_{R,D_k}}\\
&-aAH_Rm_{S_k,R}x^{-2}\left[\frac{\left(\frac{H_R}{x}\right)^a}{\left(\frac{P_{S_k}}{N_O}\right)}\right]^{-1+m_{S_k,R}}\left(\frac{H_R}{x}\right)^{-1+a}\left(\frac{P_{S_k}}{N_O}\right)^{-1}=0
\end{aligned}
\qquad (21)
$$

VI. Numerical Results

This section investigates the performance of the proposed system in terms of the outage probability for a wide range of transmit SNR values. For this purpose, a simulation setup was developed in MATLAB and the transmission of 10^6 packets for each SNR value was performed. Unless indicated otherwise, the values of model parameters used are $\left(P_{S_k}/N_O\right)=\left(P_R/N_O\right)=10$ dB, $a=3$, $H_R=20$ km, and $D_{S_kD_k}=40$ km. By adopting the recommendations for the optimal stratospheric relay location in (20) and (21), we obtain $\beta_{S_k}=\beta_{D_k}=45^\circ$. For the Rician fading conditions, it is considered that $K_{S_k,R}=K_{R,D_k}=3$ dB. Then, we obtain $m_{S_k,R}=m_{R,D_k}=1.8$.

Figure 3 and Figure 4 depict the outage probability as a function of the transmit SNR for different number of SD pairs and symmetric Rician/Rician and Rayleigh/Rayleigh fading, respectively. Both the optimal and the non-optimal location of the stratospheric relay is considered. One observes that the optimized system outperforms the non-optimized one. Moreover, a significant performance improvement can be achieved, as the number of users increases. By comparing the results in Figure 3 and Figure 4, it is also obvious that the Rician/Rician fading leads to significantly better results.

Figure 5 shows the outage probability as a function of the transmit SNR for four SD pairs and asymmetric Rayleigh/Rician and Rician/Rayleigh fading. One observes that the type of fading slightly affects the outage probability when the stratospheric relay is located at an optimal position.

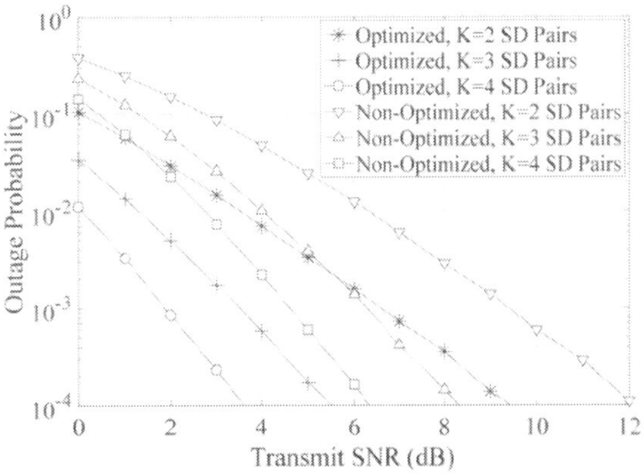

Figure 3. Outage probability of the proposed system in terms of the transmit SNR for Rician/Rician fading and different SD pairs.

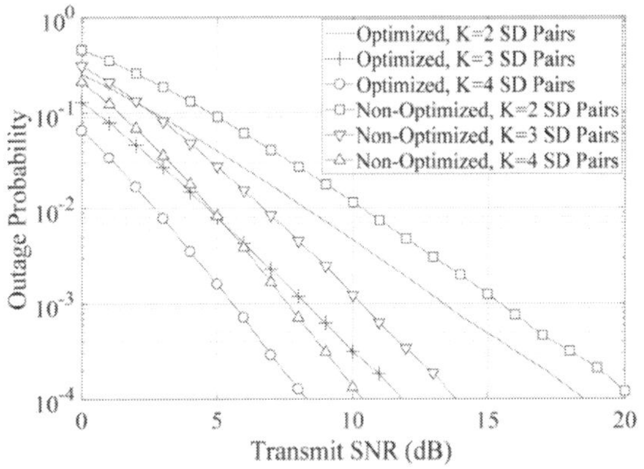

Figure 4. Outage probability of the proposed system in terms of the transmit SNR for Rayleigh/Rayleigh fading and different SD pairs.

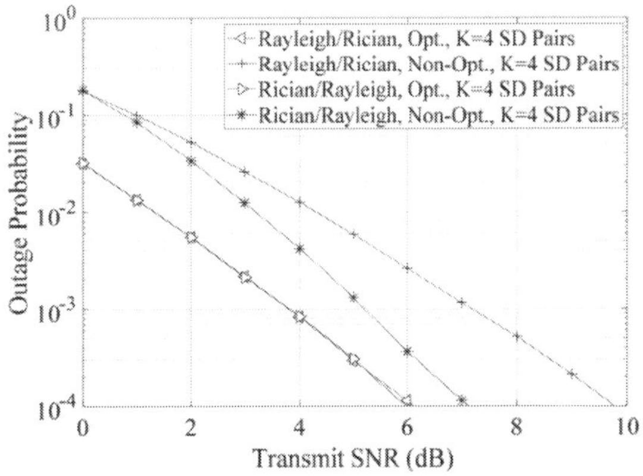

Figure 5. Outage probability of the proposed system in terms of the transmit SNR for Rayleigh/Rician and Rician/Rayleigh fading conditions.

Overall, the design recommendations ensure the successful operation and feasibility of the proposed system regardless of the variation of the propagation phenomena of both links. On the contrary, the improper position of the relay substantially influences the system performance depending on the fading conditions.

VII. CONCLUSION

In this paper, an opportunistic MS-MD pair selection scheme for stratospheric relay systems over GG fading channels has been presented. An optimum geometry-based design criterion for the placement of the stratospheric relay has been also developed, under a high SNR assumption and the impact of the number of SD pairs, the type of fading, and the proper/improper relay location on the system performance has been studied. The results have demonstrated that the performance decreases, as the fading conditions worsen, the number of SD pairs decreases, and the stratospheric relay shifts from its optimized location. These results provide guidelines for the solution of the fundamental problem of relay placement, in order to guarantee the MS-MD advantage in practice.

This work could be further improved or extended into different areas. Apart from the outage performance, the bit-error-rate (BER) and the throughput performance should be also evaluated. It would be also interesting to apply the proposed schemes to systems, where the users and the relay are equipped with multiple antennas, in order to exploit full diversity, i.e., both multi-antenna diversity and MS-MD diversity. Since power control is important in MS-MD relay networks, jointly optimizing the power allocation and the relay location is desirable. Also, in topologies where asymmetric link conditions exist, scheduling should be modified in order to ensure fairness among the SD pairs. Finally, this work can be extended to additionally support multiple stratospheric relays with free-space optical (FSO) inter-platform links.

ACKNOWLEDGMENT

This research is implemented through State Scholarships Foundation (IKY) and co-financed by the European Union (European Social Fund – ESF) and Greek national funds through the action entitled "Reinforcement of Postdoctoral Researchers" in the framework of the Operational Programme "Human Resources Development Program, Education and Lifelong Learning", with priority axis 6, 8, 9, of the National Strategic Reference Framework (NSRF) 2014-2020.

REFERENCES

[1] N. Zhang et al., "Software Defined Space-Air-Ground Integrated Vehicular Networks: Challenges and Solutions," *IEEE Commun. Magaz.*, vol. 55, no. 7, pp. 101-109, 2017.

[2] M. De Sanctis, E. Cianca, G. Araniti, I. Bisio, and R. Prasad, "Satellite Communications Supporting Internet of Remote Things," *IEEE Internet of Things Journal*, vol. 3, no. 1, pp. 113-123, Feb. 2016.

[3] A. Mohammed, A. Mehmood, F. Pavlidou, and M.Mohorcic, "The role of high-altitude platforms (HAPs) in the global wireless connectivity," *Proc. IEEE*, vol. 99, no. 11, pp. 1939–1953, Nov. 2011.

[4] J. Jiao, H. Gao, S. Wu, and Q. Zhang, "Performance Analysis of Space Information Networks with Backbone Satellite Relaying for Vehicular Networks," *Wireless Communications and Mobile Computing*, vol. 2017, Article ID 4859835, pp. 1-13, 2017. doi:10.1155/2017/4859835

[5] Q. H. Spencer, C. B. Peel, A. L. Swindlehurst, and M. Haardt, "An introduction to the multi-user MIMO downlink," *IEEE Commun. Magaz.*, vol. 42, no. 10, pp. 60-67, Oct. 2004.

[6] Y. Jeon, Y. T. Kim, M. Park, and I. Lee, "Opportunistic Scheduling for Multi-User Two-Way Relay Systems with Physical Network Coding," *IEEE Trans. on Wirel. Commun.*, vol. 11, no. 4, pp. 1290-1294, April 2012.

[7] Y. Li, M. C. Gursoy, and S. Velipasalar, "On the Throughput of Multi-Source Multi-Destination Relay Networks With Queueing Constraints," *IEEE Trans. on Wirel. Commun.*, vol. 15, no. 8, pp. 5368-5383, Aug. 2016.

[8] O. Sahin and E. Erkip, "Achievable rates for the gaussian interference relay channel," in *Proc. IEEE Global Communications Conference (GLOBECOM) 2007*, Washington, DC, USA, pp. 1627-1631, 26-30 Nov. 2007.

[9] A. Aragón-Zavala, J. L. Cuevas-Ruíz, and J. A. Delgado-Penín, *High-Altitude Platforms for Wireless Communications*, New York, USA: John Wiley & Sons, Dec. 2008.

[10] P. S. Bithas, N. C. Sagias, and P. T. Mathiopoulos, "GSC diversity receivers over generalized-gamma fading channels," *IEEE Commun. Lett.*, vol. 11, no. 12, pp. 964-966, Dec. 2007.

[11] A. J. Coulson, A. G. Williamson, and R. G. Vaughan, "Improved fading distribution for mobile radio," *Proc. Inst. Elect. Eng.-Commun.*, vol. 145, no. 3, pp. 197-202, Jun. 1998.

[12] S. S. Ikki, "Optimisation study of power allocation and relay location for amplify-and-forward systems over Nakagami-m fading channels," *Trans. Emerging Tel. Tech.*, vol. 25, no. 3, pp. 334-342, Mar. 2014.

[13] N. C. Sagias and P. T. Mathiopoulos, "Switched diversity receivers over generalized gamma fading channels," *IEEE Commun. Lett.*, vol. 9, no. 10, pp. 871-873, Oct. 2005.

[14] I. S. Gradshteyn and I. M. Ryzhik, *Table of Integrals, Series, and Products*, 6th ed. New York: Academic, 2000.

[15] S. S. Ikki and S. Aissa, "A study of optimization problem for amplify-and-forward relaying over Weibull fading channels with multiple antennas," *IEEE Commun. Lett.*, vol. 15, no. 11, pp. 1148-1151, 2011.

28 GHz Monolithic Transmitter on GaN chip for 5G application

Rajinikanth Yella
Department of Electrical Engineering
National Chiao Tung University
Hsinchu, Taiwan
rajini.02g@g2.nctu.edu.tw

Krishna Pande
Department of Electrical Engineering
National Chiao Tung University
Hsinchu, Taiwan
kppande@nctu.edu.tw

Ke Horng Chen
Department of Electrical Engineering
National Chiao Tung University
Hsinchu, Taiwan
krimaps@gmail.com

Edward Chang
International College of Semiconductor Technology
National Chiao Tung University
Hsinchu, Taiwan
edc.nctu@gmail.com

***Abstract*—5G standard is targeting much higher data rates as compared to existing wireless technologies to accommodate the ever-increasing demand for faster wireless applications. A transmitter is required to implement a 5G system. In this paper, we are presenting a 28 GHz novel monolithic transmitter architecture on GaN substrate that offers Size, Weight, Area, Power and Cost (SWAP-C) advantages. The transmitter contains a Yagi antenna, which consists of three directors, two drivers, a strip line feed, a substrate and a ground plane. According to simulation results, the designed Yagi antenna has a compact size and low loss at the selected frequency of 28 GHz. At this frequency, its return loss, gain, and beam width are -38 dB, 8.69 dB, and 57.2 degrees, respectively. The second component in the monolithic chain is a bandpass filter (BPF), which offers enhanced selectivity and stopband suppression on GaN substrate. The Bandpass filter has a minimum insertion loss of 0.6dB at 28GHz. The rejection level is higher than 10 dB in the stop band. Further, a collaborative simulation of 28 GHz mixer for upconversion with CLASS-E power amplifier (PA) with integrated octature structure to achieve robust load insensitivity is presented. In this paper to design high-efficiency PA, we implemented harmonic load pull at both the input and output of the active device to obtain optimum impedances at fundamental and second-harmonic frequencies. After an iterative process, the optimum input and output impedances are obtained. In addition, we also implemented cascaded octature power cell structure. The proposed balanced PA achieves a saturated output power (Psat) of 13.5dBm and a maximum Power Added Efficiency (PAE_{max}) of 55%. It consumes 210mW power. The presented transmitter configuration is designed on a GaN substrate with a thickness of 0.8 mm, permittivity (Er) of 9.7, and loss tangent (TanD) of 0.025. Based on its performance at 28 GHz, the designed transmitter is being developed for the 5G application.**

Keywords- Yagi antenna;Filter; Mixer;Power amplifier; 5G.

I. INTRODUCTION

Fifth generation wireless network (5G) has become research focus since it could support the explosive growth of data traffic, massively interconnected devices, and new applications. It is expected that 5G will utilize spectrum at millimeter wave (MMW) frequencies to satisfy the demand for massive bandwidth since the spectrum resources in the lower frequency bands are running out. Resources have been invested to develop prototype millimeter wave 5G mobile communication systems, especially for frequency band such as 28GHz. Therefore 27.5 to the 29.5GHz band is a strong candidate for the new 5G radio interface and much of the research undertaken to date has considered this band. For example, Samsung Electronics has built a prototype system including beamforming antenna that works at 28GHz [1], However, to our knowledge single chip transmitter on GaN does not exist for the 5G application. An overview of the architecture and components for our transmitter module on a GaN chip are shown in Figure 1. The Integrated Chip consists of a chain of various individual components such as yagi antenna, power amplifier, bandpass filter, upconverter, mixer and matching circuitry.

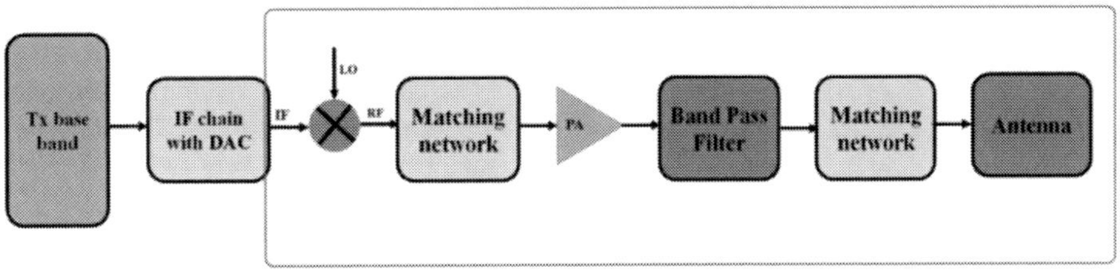

Figure 1. Transmitter architecture for 5G application

The presented transmitter configuration is designed on a GaN substrate with a thickness of 0.8 mm, permittivity (Er) of 9.7, and loss tangent (TanD) of 0.025. The goal is to ultimately fabricate the proposed transmitter module on GaN substrate.The Yagi antenna is designed using HFSS (High-Frequency Structure Simulator) simulation software. Rest of the blocks of the transmitter, bandpass filter, Mixer, Power amplifier, and matching network circuitry are designed using Advanced Design System (ADS) simulation software.

II. DESIGN AND RESULT & DISCUSSION OF TRANSMITTER MODULE ON GAN CHIP

A. *Yagi Antenna*

Yagi antenna was proposed by researchers from Japan usually for radio application [2]. Such antenna was

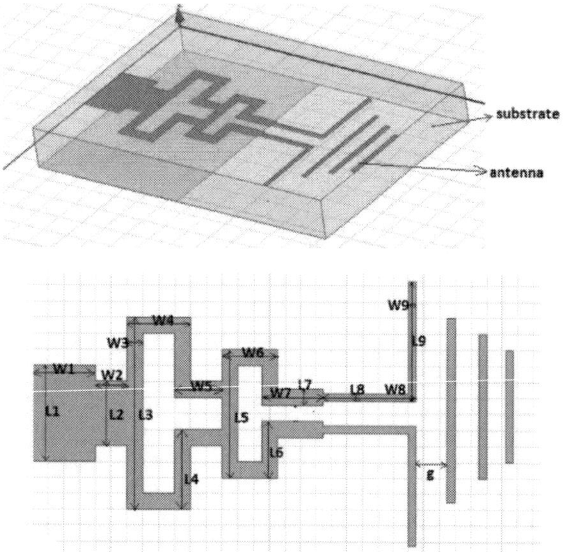

Figure 2. (a) 3D view of Antenna in HFSS (b) dimensions of yagi guda antenna

developed for X-band, Ku-band, and K-band, etc. Generally, a design of Yagi antenna places a driver and director's elements on one side of PCB and places a reflector on the reverse side. A driver is utilized for radiating electromagnetic wave whilst directors and a reflector is utilized to focus the electromagnetic wave radiation from its driver. But, the design of Yagi antenna is more challenging if it is being designed at 28 GHz on GaN substrate instead of PCB.

In this paper, we present the design of novel structure Yagi antenna, which shows good radiation efficiency and gain. The design of proposed Yagi antenna is shown in Figure. 2. It is composed of three directors, two drivers, a strip line for feed and a reflector. This antenna can be configured in MIMO format if needed. The antenna is made from gold with a thickness of 0.035 mm and electric conductivity of 4.1 x 107 Siemens/m considering GaN as substrate. In order to match antenna impedance with another device, the impedance was optimized for 50 Ohm. To fulfill such requirement, its feed- width was set to be 1.8 mm. Details of the Yagi antenna design parameters that allowed best performance is shown in Table I. Figure 3 shows simulated return loss of -38dB for 27.4 GHz to 28.4 GHz with a center frequency of 28 GHz. Antenna results are mentioned in Table 2.

TABLE I. DIMENSIONS OF PROPOSED ANTENNA

L1	L2	L3	L4	L5	L6	L7	L8	L9	g
1.8	0.9	2.5	1.2	2	0.8	0.5	0.3	2	0.1
W1	W2	W3	W4	W5	W6	W7	W8	W9	
1.5	0.7	0.5	1.5	0.5	1.5	1.8	2.5	0.3	

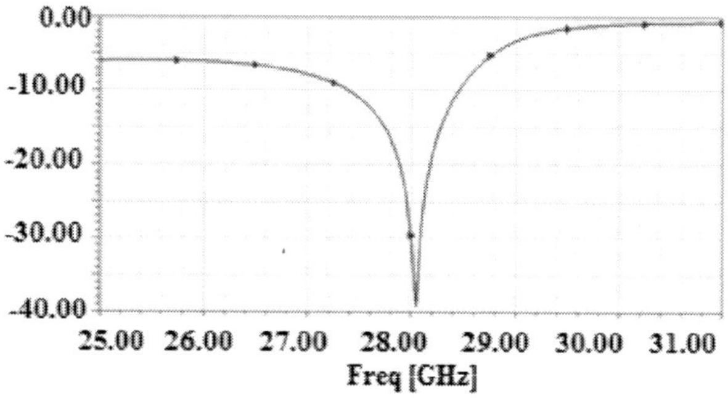

Figure 3. S11 response of return loss of antenna

TABLE II. ANTENNA RESULTS

Content		Units	
Antenna type			Yagi guda
Antenna size (L*W)		mm*mm	8*12
Peak Directivity		dB	8.9
Antenna bandwidth		GHz	27.4-28.4
Radiated power		W	0.8
Accepted power		W	0.9
Radiation Efficiency	At 28 GHz	%	96
VSWR	At 28	GHz	<1
GaN substrate details Assumed during simulation	Thickness	mm	0.8
	Er		9.7
	Tan D		0

B. *Band-Pass Filter Design*

Filters are an essential part of wireless communications systems as they are required to suppress undesired signals in the transceiver pass-band. The size, weight, cost, and loss of such filters must be kept as low as possible.

In this paper, the design of a core circuit using microstrip line has been conceived and designed as shown in Figure.4. Small size, low cost, and good performance multifunction

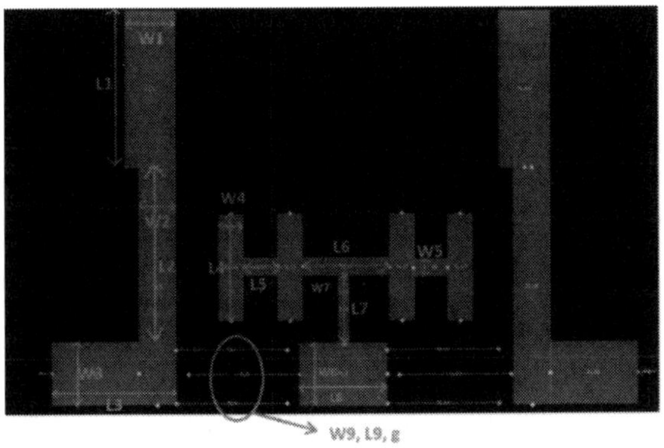

Figure 4. BPF structure with dimensions

filters are the main development trend in microwave/MMW domain. Filters with high selectivity and good stopband performances are extensively studied [3]. On the other hand, the GAN substrate has attractive characteristics, such as wide frequency range, static dielectric constant, low thermal expansion coefficient and very low water absorption [10]. In this paper, a Band-Pass Filter (BPF) based on broadside coupling H-shape resonators and half wavelength resonators are implemented on two-layer GAN substrate, which has relatively small sizes and good selectivity.

The dimensions of BPF filter are listed in Table 3.

TABLE III. DIMENSIONS OF PROPOSED FILTER

L1	L2	L3	L4	L5	L6	L7	L8	L9	g
8.8	13.5	5	6	2	5	4	5	6.5	0.5
W1	W2	W3	W4	W5	W6	W7	W8	W9	
3	2.2	3.5	1.5	1	1	0.6	3.5	0.1	

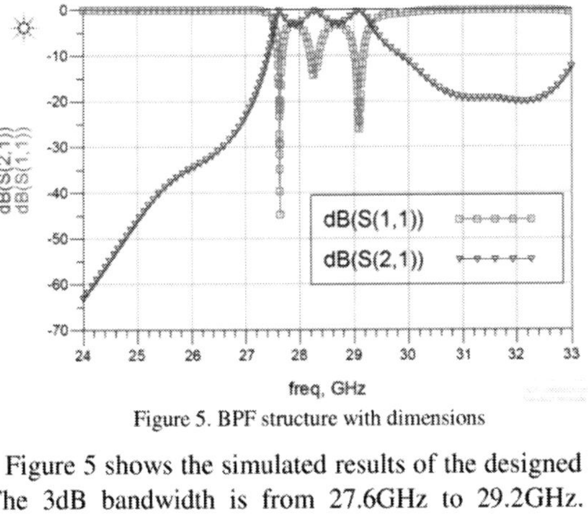

Figure 5. BPF structure with dimensions

Figure 5 shows the simulated results of the designed BPF. The 3dB bandwidth is from 27.6GHz to 29.2GHz. This filter has a minimum insertion loss of 0.6dB at 28GHz. The rejection level is higher than 10 dB in the stop band. The return loss of BPF is more than -40 dB at 28 GHz.

C. Impedance Matching Network

At 28GHz frequency, the reflected power occurs when the load impedance is not matched to the characteristic impedance of source (mixer) and load (PA). Impedance matching using the passive network is critical to achieving maximum power transfer, minimum reflection, and adequate harmonic rejection. In order to overcome such problems, we propose matching network approach is based

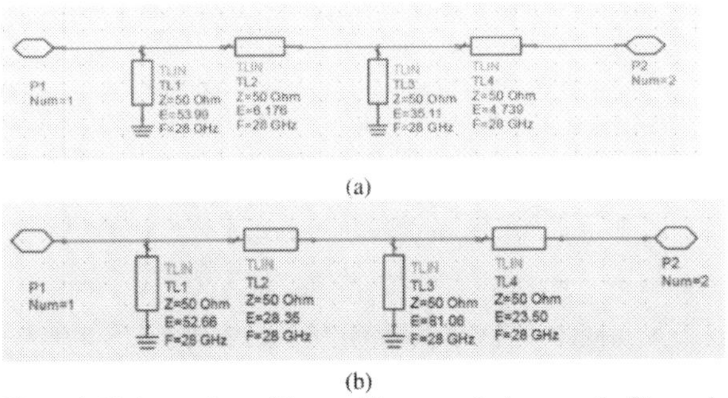

Figure 6. (a) the topology of the matching networks between the filter and antenna (c) the topology of the matching networks between the mixer and PA

on Smith chart. Through Smith Chart, we can get a matching network at a certain frequency easily. When utilizing the Smith Chart to design matching networks at the center frequency, the next steps should be followed:

- Find out the output impedance Z_{out} of previous stage transistor and the input impedance Zin of next stage transistor.

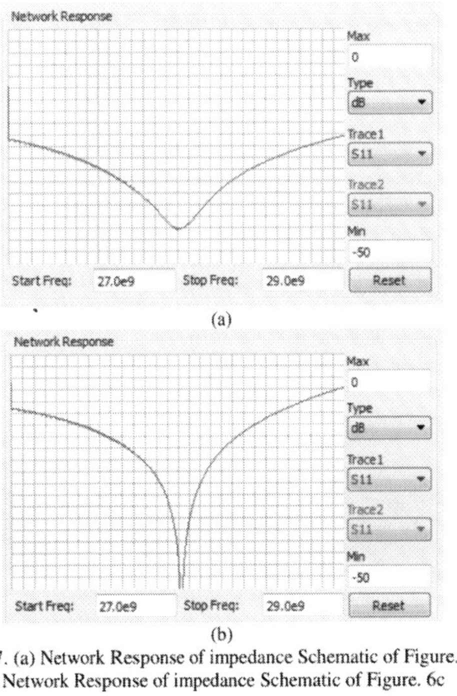

Figure 7. (a) Network Response of impedance Schematic of Figure. 6a (c) Network Response of impedance Schematic of Figure. 6c

- Find out the points of Z_{out}, Z_{in} in the Smith Chat.
- Describe the maximum equal Q curve Q_{max} from the points of Z_{ou}, Z_{in}.
- In the extension of the Smith Chart constructed by Q_{max}, add appropriate microstrips and capacitances to make Z_{out} and Zin conjugate match.

Then, consider the network composed of micro-strips as the initial matching network. We designed such matching network using ADS. This network is designed to operate at 28 GHz frequency with a normalized impedance of 50 Ω. Where ZS is source output impedance, ZL is load input impedance. We calculated stub values mentioned in Figure. 6 using ADS. Figure. 6 shows the objective function to optimize the whole matching networks. Figure.7 shows return loss plots of machining networks; the rejection level is higher than 10 dB in the stop band.

D. *Mixer Design*

The transmit Signal coming from baseband is almost at DC, hence it must be upconverted to high frequency at which antenna is working. In order to do that signal must be up-converted using MMW mixer. Doubly-balanced Mixers (DBM) are usually the desirable mixer because of their superior suppression of spurious mixing products and good port-to-port isolation [4]. An ultra-wideband balanced microstrip balun was reported [4], but this type of balun was difficult to fabricate. A 28 GHz mixer based on the Marchand balun was fabricated in 0.15ȝm GaAs PHEMT technology, similar to our proposed design in GaN technology.

In this paper, a 28 GHz doubly-balance mixer based on the [12] Marchand balun in the 27GHz to 30 GHz RF/LO range and DC to 5GHz IF is presented. The collaborative simulation of HFSS and ADS has been adopted in mixer design, the balun in LO port and RF port are simulated by the full-wave electromagnetic simulator in HFSS, then the S parameter data of the balun are exported to the mixer circuit in ADS. The mixer offers about 10dB typical conversion loss, high gain compression, higher than 20dB LO-to-RF isolations and about 10dB return loss across 27GHz to 30 GHz.

1) *The Balun Design*

The balun was simulated by using Ansoft HFSS and Agilent ADS. The Marchand balun was compensated by two open-circuited stubs at the output ports, the coupled line model in the circuit initially was used to predict the performance, and then simulated by using the full-wave electromagnetic simulator to improve the accuracy of the simulation finally. The 3D structure of the balun used in the mixer was drawn by HFSS and shown in Figure.8, which is designed for GaN substrate with thickness 0.8mm and dielectric constant 9.7. 1 is the input port and 2 3 are output ports. Impedances of three ports are all set to be 50 Ω and the microstrip line width is 0.25mm, and l. w2 and w3 stand for the length the width and the gap width of the microstrip lines, the hole is metallization and connected to ground. The Marchand balun is simulated by optimizing w2 and w3.

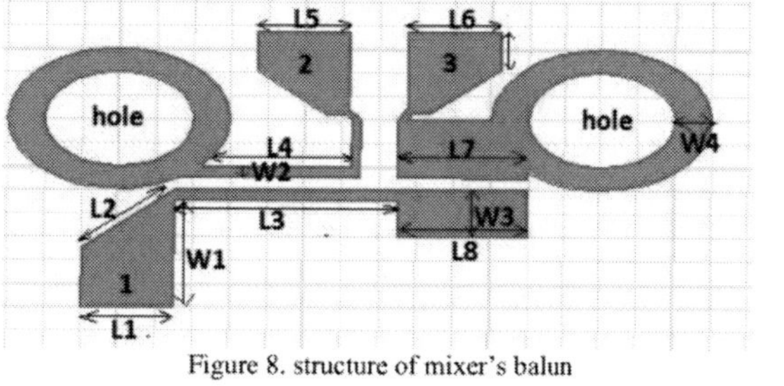

Figure 8. structure of mixer's balun

TABLE IV. TABLE TYPE STYLES

L1	L2	L3	L4	L5	L6	L7	L8
1.2	1.5	2.5	1.8	1.2	1.2	1.4	1.4
W1	W2	W3	W4				
1.2	0.2	0.5	0.3				

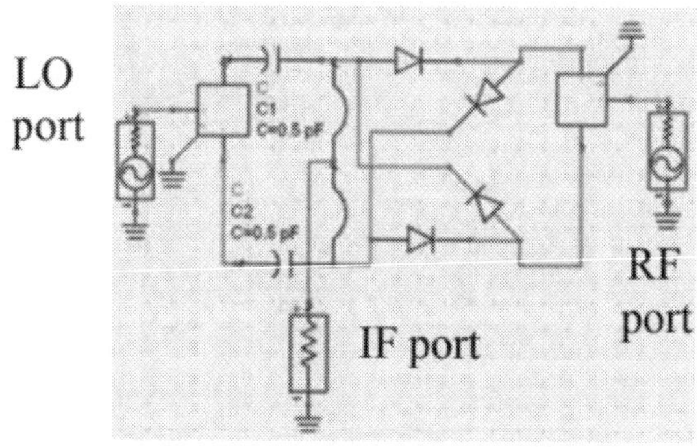

Figure 9. Double balanced mixer circuit in ADS

2) *The Mixer Design*

Then, the terminal S parameter data of the balun in HFSS are exported to the mixer circuit in ADS in Figure.9, the three ports network are set up to substitute the balun. RF

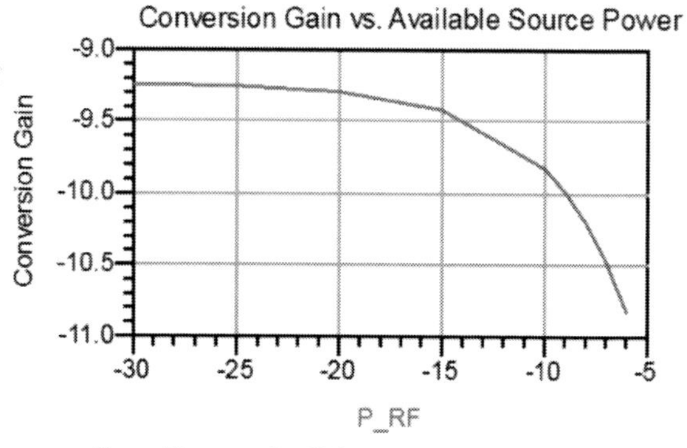

Figure 10. conversion Gain vs available source power

port and LO port is power source port, the IF port is load port, all the port is set to 50Ω. Two gold lines are set to near the real condition, the line is set to 5mm length and the diameter is 0. 1mm. The capacitances are used in LO port to make the minimal frequency of the IF port extend to DC. C1=C2=0.5Pf. The simulated result of the mixer circuit conversion gain was obtained using the harmonic balance method in Figure. 10. The type conversion gain is about 9. 3dB.

E. Power Amplifier Design

For wireless communication systems, power amplifiers (PAs) with high efficiency is critical as they consume the majority of the power of the systems and closely related to the thermal problem. As a result, many categories of high-efficiency PAs have been widely studied in recent years The Class-E amplifier is well known for its high efficiency and simple structure. Moreover, inherent output capacitor of a transistor deteriorates the performance of a Class-E amplifier at high frequency. To address above problems, the Class-E amplifier was designed by modifying the device output impedances at odd and even harmonics, so as to shape the output voltage and current that minimize the overlap to achieve high efficiency [7].

In this paper, a simple method of designing high-efficiency PA is presented. Compared with other works, this method is easier to design harmonic tuned PA and thus achieves relatively high efficiency with fewer harmonics to control. The simulated results indicate a high-efficiency Class-E PA is realized with 55% PAE and 13.5 dBm output power at 28 GHz.

However, the challenge of the MMW PA design is to deliver maximum output power with a maximum Power Added Efficiency (PAE) and high linearity [8]. For this purpose, to achieve high efficiency, we have implemented two techniques. Technique one is harmonic load pull conducted at both the input and output of the active device to obtain optimum impedances at fundamental and second-harmonic frequencies. After an iterative process, the optimum input and output impedances are obtained. Technique two is cascaded octature power cells structure is implemented. The presented circuit is implemented in GaN HEMT technology taking, the advantage of high-frequency performance and high power performance [9].

1) Design process:

To achieve the target output power, a total of eight stages were power-combined in the output stage. This output stage was driven by a pair of devices, and this, in turn, was driven by an input stage realized using a single transistor. The transistor sizes in each stage needed to be identical, and the basic power-combining topology needed to be similar.

The individual PA structure is shown in Figure. 11 is composed of three series transmission lines with one parallel open-circuit stub located between the first two lines are used as the impedance matching circuit, which is a simplified method of the distributed multi-frequency matching approach, as introduced previously [10]. The single transmission line before the open stub is used to tune the impedance of the second harmonic, while the two series

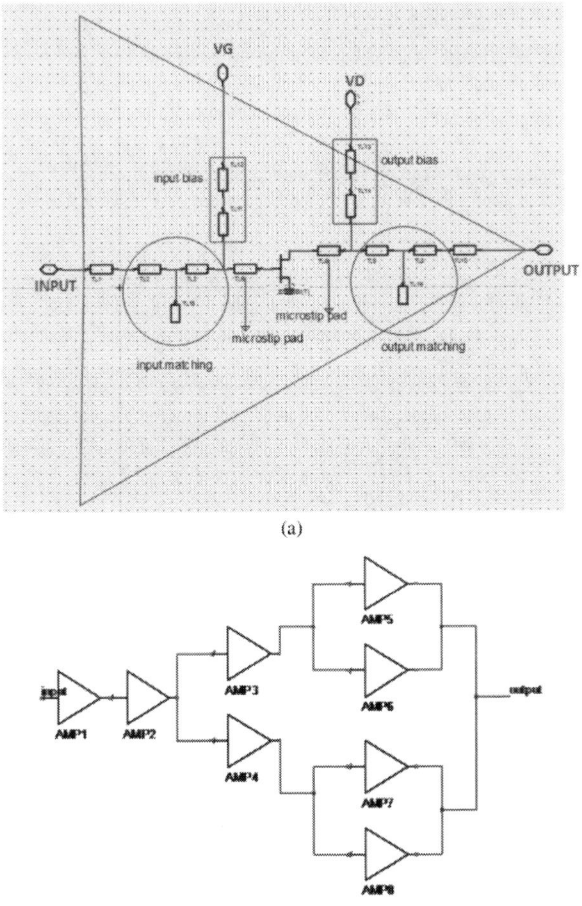

Figure 11. (a) single stage PA design (b) octature structure PA design

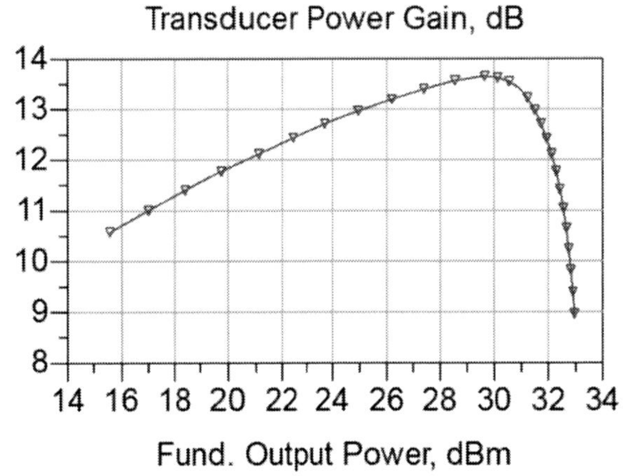

Figure 12. Transducer power gain of PA for single stage

transmission lines after the open stub are used to tune the impedance of the fundamental mode. Therefore, optimum impedance control at fundamental and second-harmonic frequencies can be realized simultaneously. The simulated large-signal performance is plotted in Figure. 12.

F. End to End Module performance

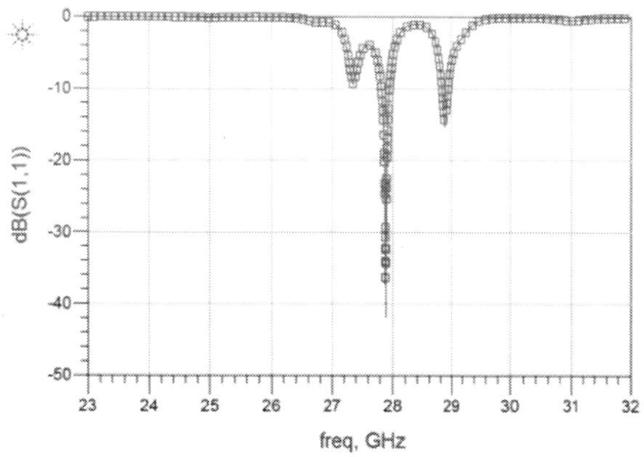

Figure 13. Simulated transmitter architecture end to end S11 response

To investigate the loss of proposed transmitter architecture, we have simulated S11 of transmitter module chain (with Yagi antenna, a BPF, mixer, matching network and power amplifier) in ADS. The results are shown in Figure.13. We plotted S11 to find the loss of the whole module and the S11 value at 28 GHz is lower than -30 dB, which is excellent.

III. Conclusion

The design of a novel transmitter module on a single chip using GaN/Si as the substrate is presented in this paper. The transmitter module consists of Yagi antenna, BPF, impedance matching network, up-conversion mixer and power amplifier. We investigated innovative designs for the transmitter components with potential for integration on a GaN substrate. To validate our designs, we selected 28 GHz but module design can be easily scaled to higher millimeter wave frequencies, such as 60 GHz for the 5G system. Designed transmitter module shows loss of less than -10 dB at 28 GHz, which can be further improved with design refinement. To our knowledge, this is first reported the design of monolithic transmitter module on GaN chip. Our proposed module will offer significant SWAP-C (size, weight, area, power and cost) advantages.

Acknowledgment

We are thankful to our National Chiao Tung University, Taiwan for providing the facilities to carry out our research. The project work, described in this paper, is intended for thesis for students Yella who is also grateful to colleagues in EE Dept.

References

[1] T. Kim, J. Park, J.Y. Seol, S. Jeong, J. Cho and W. Roh, "Tens of Gbps support with mmWave beamforming systems for next-generation communications," IEEE Global Communications Conference (GLOBECOM), Atlanta, GA, 2013, pp. 3685-3690.

[2] G. Sato, "A secret story about the Yagi antenna", IEEE Antennas and Propagation Magazine; Jun. 1991, vol. 33, pp. 7-18.

[3] D. Liu, U. Pfeiffer, J. Grzyb, and B. Gaucher, "Advanced Millimeter-wave Technologies," John Wiley and Sons, 2009.

[4] R. Giofre, P. Colantonio, F. Giannini , and L. Piazzon, "A new design strategy for multi frequencies passive matching networks" 2007 Eur. Microw. Conf., IEEE; 2007, p. 838–41. doi:10.1109/EUMC.2007.4405323.

[5] Z.P. Li, and J. Liu, "Design of a Novel Ultra Wide Band Balance Microstrip Balun", Information Science and Control Engineering (ICISCE), 2015 2nd International Conference on. IEEE, 2015: pp. 694-696.

[6] J. Kampa and K. Petrus, "Microwave amplitude equalizer", Microwaves, Radar and Wireless Communications. 2000. MIKON-2000. 13th International Conference on. IEEE, 2000, pp. 37-40.

[7] L. Zhongpu, D. Xiu, W. Liutai, S. Jiawen, and L. Jinxian, "Design of A Ka-band double balance mixer based on the Marchand balun", Science and Technology on Electronic Test & Measurement Laboratory.

[8] S. Shakib, H. C. Park, J. Dunworth, V. Aparin, and K. Entesari, "A 28GHz efficient linear power amplifier for 5G phased arrays in 28nm bulk CMOS," in 2016 IEEE International Solid-State Circuits Conference (ISSCC), 2016, pp. 352–353.

[9] M. N. Ruiz , D. Vegas , J. R. Pérez-Cisneros, and J. A. García, "GaN HEMT Class-E Rectifier for DC+AC Power Recovery" Department of Communications Engineering, University of Cantabria, 39005 Santander, Spain.

[10] P. Saad, P. Colantonio, L. Piazzon, F. Giannini, K. Andersson, and C. Fager, "Design of a concurrent dual-band 1.8–2.4-GHz GaN-HEMT Doherty power amplifier," IEEE Trans. Microw. Theory Techn., vol. 60, no. 6, pp. 1840–1849, Jun. 2012.

Leveraging Commercial Software-Defined Radio for Low Cost Deep Space Testing

Timothy Pham, Leslie White
Jet Propulsion Laboratory
California Institute of Technology
Pasadena, California, USA

e-mails: {Timothy.Pham, Leslie.White}@jpl.nasa.gov

Abstract – **In a typical space mission development life cycle, there is a stage where the spacecraft needs to test against the ground station for interface compatibility to ensure that the spacecraft will be properly tracked after launch. This testing normally requires the spacecraft team to bring their flight equipment to the ground station facility. While recognizing that testing with actual flight or engineering module is the most preferred option because of maximum fidelity, there are occasions when the use of actual flight hardware is a logistical challenge because of constraints in spacecraft development schedule. Having another test tool that can emulate the spacecraft signal – by recording the signal transmitted by the spacecraft and later regenerating a radio frequency (RF) signal for ground system testing - would be very useful. It is even more beneficial if such spacecraft emulator is inexpensive and highly portable. In this paper, we describe a new approach that enables ground station testing in support of deep space missions. This is very different from the traditional approach used in the Deep Space Network (DSN) where test equipment is custom designed and built, with much more capability in generating signal of different characteristics, but also at a much greater cost. The new approach, which can be used in complement with the traditional DSN test signal generator or with spacecraft flight radio equipment. The new test capability involves a very low-cost and light-weight recorder/playback assembly (RPA). This equipment leverages on the commercially available Software-defined Radio (SDR) and public-domain software. The RPA, thus far, has been used to support two missions under the National Aeronautics and Space Administration (NASA). One effort is to validate that the Uchinoura 34-m tracking station of the Japanese Aerospace Exploration Agency (JAXA) can track the upcoming NASA Exploration Mission 1 (EM-1) spacecraft, scheduled for launch in 2019. The RPA helps the EM-1 team keeps their flight system for other development/testing needs. The second effort is to help with the testing certification of the 21-m antenna ground station at the Morehead State University (MSU) in Kentucky, United States, prior to the time when the NASA Lunar IceCube spacecraft is ready for actual compatibility testing. The RPA also enables MSU student/staff training on the operations of the new ground station. This low-cost test signal generator allows the MSU team to save money and effort by not having to develop a full-scale self-generated telemetry test signal source.**

Keywords - SDR; RF Test Capability; Uchinoura; Morehead State University Ground Station.

I. Introduction

In a typical life cycle of deep space mission development, there is a period prior to spacecraft launch when the flight communications system needs to be tested for interface compatibility against the ground station that will be later tracking the spacecraft. The compatibility testing normally requires the spacecraft team to bring their flight radio to the ground station test facility. The tests would exercise various modes of spacecraft operation, e.g., different data rates, coding schemes, signal levels, to ensure that the signal can be properly processed by the ground station. Under the philosophy of "test as you fly", this is a good practice to ensure the compatibility between flight and ground system; however, there are occasions when such a test poses logistic problem to the flight team, e.g., their equipment may be tied up with other development effort. Having a test instrument that can serve as a substitute for flight equipment – a tool that records the signal transmitted by spacecraft and recreates an RF signal for injection into the ground station under test - would offer more flexibility. The benefits are even greater if such a spacecraft emulator is very affordable and highly portable, as in this case, because it reduces the financial burden on the mission and simplifies the shipping logistics when conducting tests.

Section II of this paper discusses a design and capability of the Recorder/Playback Assembly (RPA), developed at the Jet Propulsion Laboratory in support of NASA missions. The benefits of the equipment are captured in Section III. Some possible future improvements to the equipment operations, gained from our testing experiences, are offered in Section IV. Support to the EM-1 mission testing is described in Section V, followed by a similar discussion on test support at the Morehead State University in Section VI.

II. Design Architecture

The Recorder/Playback Assembly offers a complementary capability to the test equipment typically developed in the NASA Deep Space Network. Up to now, the test equipment is custom designed and built to the specifications so it can generate test signals that emulate various characteristics of spacecraft that require DSN support. Such development requires significant effort, both in terms of financial resource and time. The new SDR-based equipment described in this paper leverages on commercial products; thus, enables a quick development at low cost. The RPA can be used in complement with the DSN test signal generator by recording its signal and replicating at another facility; thus, enables the test capability at the new site without much cost and effort.

The RPA, as seen in Figure 1, comprises of three key components: (1) a laptop computer with a very fast input/output (via USB3 connections) to enable data transfer from the transceiver to a disk storage, (2) a high capacity data storage (~3 Terabytes or greater), and (3) a commercial SDR RF transceiver. The computer provides the graphical user interface for user to control the signal generation or signal recording. It configures and controls the setting of the transceiver. In the recording mode, the computer transfers a 10-MHz digitized samples from the transceiver to the disk storage. In the playback mode, it reverses the digital sample flow, from the disk storage to the transceiver. In addition, the laptop also generates a Fast Fourier Transform (FFT) spectrum of the recorded or playback signal, using the digitized samples. The second component – the disk storage - archives the samples. Both the computer and disk storage use a high speed USB3 interface for data transfer. For a 10 MHz I/Q sampling, the data throughput can be as much as 80 MB/s.

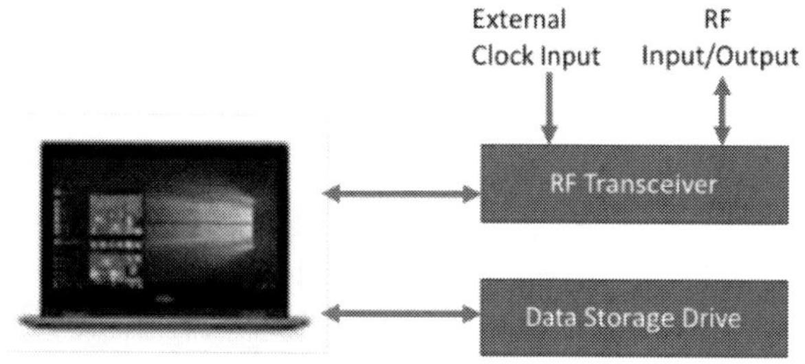

Figure 1. Components of the Recorder Playback Assembly

The most crucial element of the RPA is the RF transceiver [1]. On the recording, it is equivalent to a heterodyne receiver where the RF signal is first down-converted to an IF frequency and then digitized. On the playback, the functions are reversed. The particular commercial unit we employed in the RPA can digitize an RF signal input up to a rate of 20 Msps, I/Q sampling. To enable the recorded data possibly used by other SDRs (e.g., for future upgrade), we chose to have the digitized samples written in the standard 32-bit format. This prompts a high I/O rate (80 MB/s) and a large data storage capacity. Our particular chosen SDR supports an input/output RF frequency range of 1 MHz - 6 GHz. This allows the signal recording, or playback, directly at S-band frequency (~2.3 GHz) that the ground station expects to be receiving from the EM-1 spacecraft. For the Morehead State University testing with the Lunar IceCube spacecraft, since the spacecraft RF transmitted signal will be at X-band (8.4 GHz) which is beyond the capability of the transceiver, the recorded and playback signal are done at an intermediate frequency (IF) around 300 MHz. In the ground system under test, this signal injection point is after the first RF/IF downconversion and prior to the carrier and symbol demodulation.

The SDR transceiver can operate with an internal clock or be synchronized to an external 10 MHz reference. Since the SDR expects a Low Voltage Transistor Transistor Logic (LVTTL) input level for the frequency reference, we needed to build an adapter to convert a typical sinusoidal reference (+13 dBm) to that of LVTTL. This adapter essentially is a resistor-capacitor-diode network to level shift and it clamps the signal level to the 0-volt minimum and 3.3-volt maximum, as required for LVTTL compatibility. The use of an external frequency reference from a highly stable clock, such as those produced by a Hydrogen maser at both ground stations of JAXA and MSU, produces a significantly more stable signal, which is critically important to our test objectives. Performance of the signal stability with and without the external reference is later discussed in section VI.

The transceiver control software is leveraged on public-domain software available from the gnu library. This is another benefit for using the commercial SDR. We developed two modes of operation, either recording or playback. Figure 2 shows a sample of the graphical control interface for data recording. By specifying the local oscillator frequency, typically a few MHz off from the actual carrier frequency, we can position the signal to be sampled away from the dc component. Figure 3 shows a similar control window for the playback where one can specify the selection of the data file and output frequency, along with other detailed configuration parameters such as the setting of RF, IF and baseband gain. In both recording and play back modes, an FFT spectrum, as seen in Figure 4, is provided to aid with signal monitoring.

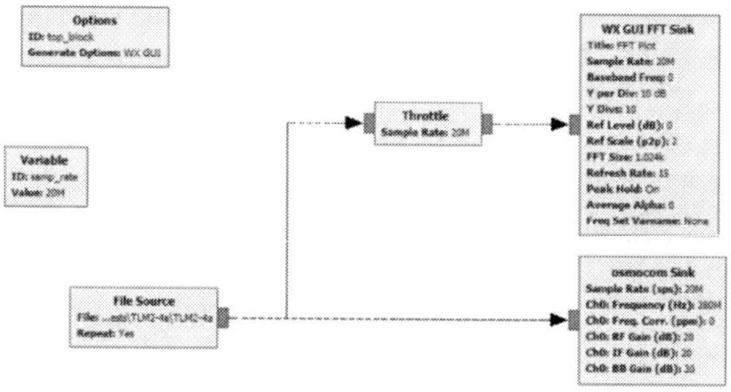

Figure 2. Control Interface for Recording

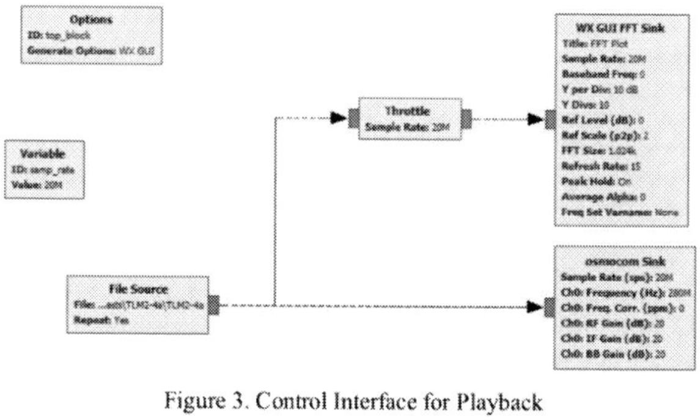

Figure 3. Control Interface for Playback

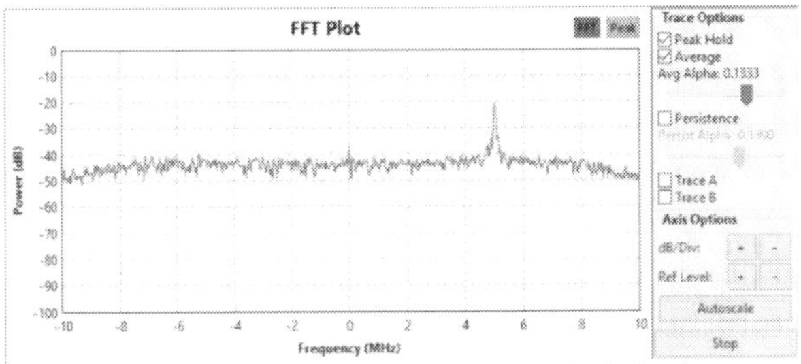

Figure 4. Spectral monitor of recorded/playback signal

These control interface windows make it easy to operate the system. Typically, only a few parameters such as the input/output filename and the signal frequency need to be changed when running different tests.

III. BENEFITS

From our experience, below are some key benefits of the RPA.

(1) *Low Cost* - In application where one needs to replicate a test signal source that is only available from other high-end equipment, this low-cost recorder/playback capability is an attractive option. The hardware cost, purchased in 2016, is around $2K. About 1 month of effort is required to assemble and test the monitor & control software, leveraging on the programming codes available to the public.

(2) *Portability* - Another feature we highly value is the equipment portability. The whole RPA weighs just a bit heavier than a common laptop, less than 7 lbs., making it very easy to hand carry to a distant test site.

(3) *Upgradability* - Since the SDRs are commercial products and supported by a large user community who also provide many standard interfaces and functionality, as more powerful products become available, the system could be easily upgraded without much effort.

(4) *Flexibility* - The wide range of frequency supported by the transceiver provides much flexibility with testing. It allows us to inject the test signal directly into the front-end of ground system at S-band frequency (the same signal path traveled by spacecraft signal during operational phase), or at some intermediate frequency in mid-stream of the system. The ease of setting the carrier frequency makes it easy to simulate different spacecraft signals under test.

(5) *Efficiency* - The RPA can generate test signal with the data-repeat option. It can continually regenerate the signal with only a small set of recorded data; thus, reduces the need for a large data storage or a long recording session.

IV. FUTURE IMPROVEMENTS

Although the RPA sufficiently supports our testing needs, there are certain features we could pursue in the future that would improve its operability:

(1) *Gain setting* - The gnu library allows for the gain setting of three stages: baseband, IF and RF. By setting these gains in midrange, we are able to capture and regenerate the signal at nominal power level and with good quality (i.e., without distortion). Having a better understanding on how various combinations of gain setting work would help toward creating a signal of different power levels that is best suited for various operating conditions.

(2) *Trimming of recorded data* – Some of our recorded data files started earlier than the actual data segment of interest. Under the current capability, the entire data file is replayed, resulted in some wait time before reaching the segment of interest. Having utility to trim the data file to just segments of interest would increase the efficiency of testing.

(3) *Higher frequency* – As more commercial products become available with greater capability, we would consider upgrade the SDR to those operating at X-band (8.4 GHz) and beyond. This would further enhance testing capability for X-band, as well as Ka-band (26-32 GHz), missions.

V. EM-1 MISSION SUPPORT

One objective of the RPA development is to provide test support to the Exploration Mission 1, currently scheduled to be launched in late 2019 [2]. The EM-1 spacecraft will be on a three-week trajectory that takes it to the Moon, stays in the parking orbit for a few days, and then returns to Earth. EM-1 is intended to demonstrate the operation of a new spacecraft system, along with the new Space Launch System, prior to the Exploration Mission 2 that will be carrying the astronaut crew. In support of future crewed mission, precise navigation and spacecraft orbit determination, especially for the Earth return segment, are important. For EM-1, most of the tracking during the mission three-week operations will be provided by the NASA Deep Space Network; however, in the interest of getting supplement tracking data during the critical phase of the mission, e.g., Earth return, additional 3-way Doppler data from the JAXA antenna is deemed beneficial to mission navigation.

The focus of JAXA support is mainly on tracking EM- 1 carrier signal and providing 3-way Doppler data to the mission navigation. 3-way refers to a mode of operation where the receiving antenna tracks the spacecraft signal is coherent to the uplink of another transmitting antenna. In our case, the DSN antenna would provide an uplink to the spacecraft and JAXA antenna will track the spacecraft downlink, in concurrent with the DSN tracking. No telemetry data processing is needed from JAXA Uchinoura station.

Normally, we could leverage a spacecraft currently in flight that has the same signal format to test the ground system compatibility with future spacecraft. However, EM-1 signal format is different from other lunar missions currently in operation, such as the Lunar Reconnaissance Orbiter (LRO). EM-1 telemetry data is directly modulated onto the carrier. In contrast, the LRO telemetry data is first modulated on a subcarrier, prior to the carrier modulation. To ensure that Uchinoura system can track the EM-1 signal, it was decided that a replica of EM-1 signal – being recorded and played back by the RPA – should be used to check out the compatibility of the flight and ground interface. The main objective is to demonstrate that the receiver at the Uchinoura station can track the carrier in the presence of telemetry modulation, and to provide good, stable Doppler data.

A preliminary test was recently conducted using the RPA to characterize the Doppler data tracked by Uchinoura station [3]. The RPA previously recorded the EM-1 RF test signal during its compatibility testing with the Deep Space Network. The recorded data were then used to generate a duplicated EM-1 signal at 2.3 GHz (S-band). The RPA output was injected into the front end of the Uchinoura station, in front of the low noise amplifier. The Uchinoura receiver successfully locked to the RPA signal and produced Doppler data (measured carrier frequency) as shown in Figure 5. Testing was done for several EM-1 configurations of different telemetry data rate ranging from 72 ksps to 2 Msps. The test also demonstrated that it could track EM-1 signal with both the Costa loop (relying on telemetry symbols only, not carrier) and the standard phase lock loop (relying on the residual carrier buried underneath the telemetry spectrum).

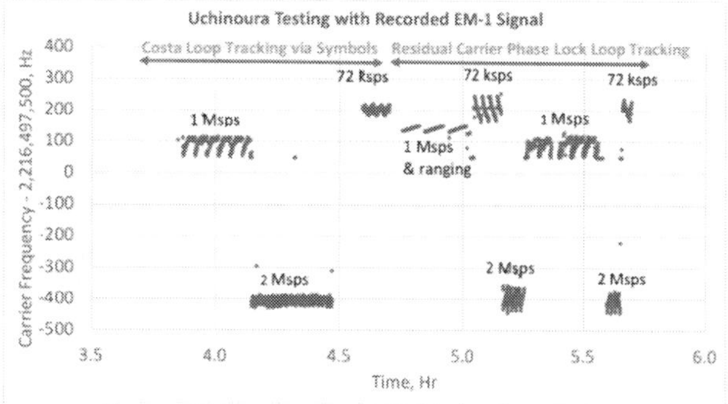

Figure 5. Uchinoura test with EM-1 signal, 2017-12-01

One may observe that there is a cyclic pattern within each data rate configuration. This is caused by the data repeat feature in the play back mode discussed earlier in Section III.

VI. MSU SUPPORT

The 21-m antenna system at the Morehead State University is being upgraded to support the Lunar IceCube mission, and likely also for other CubeSats launched on the EM-1 mission along with the Lunar IceCube [4][5]. This system will be capable of deep space tracking at X-band (8.4 GHz). The system has some DSN-replica digital equipment that are specialized in telemetry, tracking and command in deep space environment. Due to limited budget and time constraint, the 21-m system does not have a full-scale test equipment that can simulate the Lunar Ice Cube signal. The RPA helps to fill this gap. It first recorded a test signal of Lunar IceCube characteristics that was generated by a more capable DSN Test Signal Generator. The signal was then recreated and injected into the MSU ground station. Through this effort, the RPA helps verifying the ground system components and building up the confidence that the MSU ground system would be ready to support further interface testing with the actual Lunar IceCube flight system when it becomes available.

The recorded Lunar IceCube test signal comprises of a suppressed carrier modulated by telemetry data which are encoded with Turbo code, rate 1/6. The preliminary test results demonstrated that the MSU equipment can successfully demodulate and decode the Lunar IceCube telemetry data at 64 ksps. Further testing is needed to demonstrate similar performance at 384 ksps where Lunar Ice Cube is expected to operate.

During the MSU testing, we were also able to characterize the performance of the RPA with and without the 10 MHz reference obtained from the on-site hydrogen maser clock. Figure 6 shows the measured frequency stability of the test signal, as detected by the receiver. With the 10 MHz reference input, the carrier signal was very stable. The frequency variation was within 20 mHz over short term and 70 mHz over a period of 2.5 hrs.

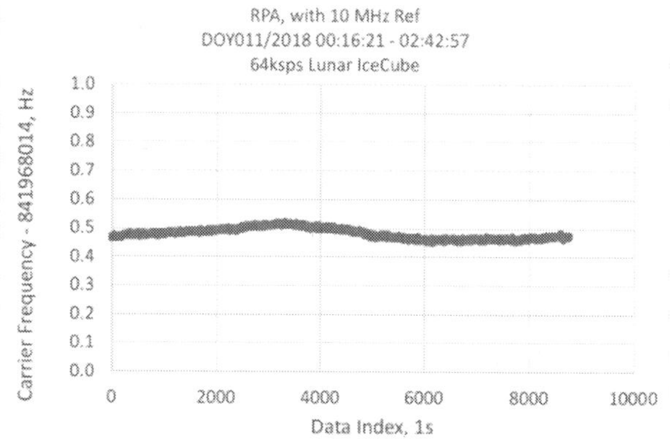

Figure 6. Stability of the test signal, with external reference.

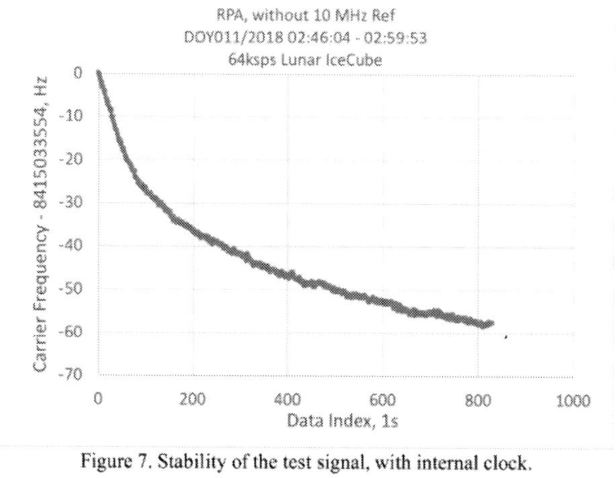

Figure 7. Stability of the test signal, with internal clock.

When generated with the transceiver internal clock, the carrier frequency drifted as much as 60 Hz just over 14 minutes, as shown in Figure 7. With such variation, while it is possible to track the carrier with a wider bandwidth, it would be difficult to maintain symbol demodulation with the maximum bandwidth allowed by the receiver. Our assessment is that in order for the RPA to support telemetry data testing, it needs to be synchronized to a stable external frequency reference.

VII. Conclusion

In this paper, we discussed the use of a commercial software-defined radio in our testing effort at the JAXA Uchinoura and the Morehead State University ground stations. This test tool allows us to record, and later play back, the EM-1 signal generated by the spacecraft under development, and the emulated Lunar IceCube generated by the DSN Test Equipment. The test capability is of low cost, both in term of hardware procurement and software development, thanks to the available commercial products and public domain SDR software. One most valuable feature is the portability of the system. Its light weight makes it very easy to transport and conduct test at other distant ground station, without a need for transportation shipment or on-site installation.

Both preliminary testing at Uchinoura and MSU have been successful. At Uchinoura, the ground station demonstrated it could track the recorded EM-1 signal. At the Morehead State University, the RPA enabled demonstration of functionality of the ground station, where telemetry data emulating the Lunar IceCube mission was successfully demodulated and decoded. From our testing, we learned that it is critical to have the RPA using an external atomic frequency reference – instead of relying on the internal clock - in order to achieve the necessary frequency stability needed for Doppler and telemetry processing.

Acknowledgment

The work described in this paper was carried out by the Jet Propulsion Laboratory, California Institute of Technology, under a contract with the National Aeronautics and Space Administration. We would like to thank Hiroshi Takeuchi and Atsushi Tomiki at JAXA for their indispensable assistance with the Uchinoura testing. Support in conducting additional tests at the Morehead State University by Sarah Wilczewski and data extraction by Jason Liao are much appreciated.

References

[1] HackRF One, https://greatscottgadgets.com/hackrf/, retrieved: Mar. 2018.

[2] EO Sharing Earth Observation Resources, eoPortal Directory, Orion/EM-1(Exploration Mission 1), https://directory.eoportal.org/web/eoportal/satellite-missions/o/orion-em-1, retrieved: Mar. 2018.

[3] Japan Aerospace Exploration Agency, Uchinoura Space Center, http://global.jaxa.jp/about/centers/usc/index.html, retrieved: Mar. 2018.

[4] P.E. Clark et al., Lunar Ice Cube Mission: Determining Lunar Water Dynamics with a First Generation Deep Space Cubesat, 47th Lunar and Planetary Science Conference (2016), https://www.hou.usra.edu/meetings/lpsc2016/pdf/1043.pdf, retrieved: Mar. 2018.

[5] J. Kruth, B. Malphrus, J. Wyatt, and T. Pham, "Enabling University-Operated Ground Support for Deep Space Small Spacecraft Missions – A Pilot Development with the Morehead State University 21-m Ground Station", 2017 Interplanetary Small Satellite Conference, San Jose, CA, May 1-2, 2017, https://www.dropbox.com/sh/g9ran9t1itphckk/AADVVbRWHhw8XOOGEZsQvpYwa/C2-J.%20Wyatt?dl=0&preview=Enabling+University-Operated+DSN+Station+v.4.28.2017.pptx, retrieved: Mar. 2018.

Web-based Geographical Information System for Real-Time Flood Monitoring of the River Arachthos in Epirus Region, Greece

Stavros Kolios[1,2]
[1]Department of Computer Engineering, Technological and Educational Institute of Epirus
Arta, Greece
[2]Department of Geography, University of the Aegean, Mytilene, Greece
e-mail: stavroko@gmail.com

Petros Karvelis[1]
[1]Department of Computer Engineering, Technological and Educational Institute of Epirus
Arta, Greece
e-mail: pkarvelis@teiep.gr

Chrysostomos Stylios[1]
[1]Department of Computer Engineering, Technological and Educational Institute of Epirus
Arta, Greece
e-mail: stylios@teiep.gr

Abstract— **The scope of this work is to present the design and the developmental stages of a Web–based GIS (Geographic Information System) for flood monitoring of the Arachthos river in the Region of Epirus. The system is designed to cover the specific river characteristics as well as the potential flood affection in the greater area along the riversides. The development of the WebGIS is based on the client–server model and uses Google Maps API (Application Programming Interface) services for data plotting. Specific scientific instrumentation for monitoring the height of the river water and the current meteorological conditions is connected automatically to the system, providing relative measurements in real-time basis. Modern methodologies are used so that to provide valuable information and to create a trustful and accurate system for real-time monitoring of river flow and the early warning of possible floods.**

Keywords- river floods; early-warning; monitoring; WebGIS.

I. INTRODUCTION

Natural hazard of floods, as well as its risk assessment is a major topic of interest among many scientific communities and the local/national governance, worldwide [1]. The main reason behind this global interest is the increase of storms in terms of frequency, magnitude and impact, which is partly owed to the climate change [2]. Therefore, it is important to analyze where these hazard events can occur, with what frequency, and which is the vulnerability of an area in cases of possible flooding and/or the river flow expansion beyond its riversides. The real-time monitoring, as well as the risk management and assessment includes suitable instrumentation, and automated methodologies for the estimation of the level of risk, followed by an evaluation of this risk level [3][4].

Geographical Information Systems (GIS) are already playing a major role in the flood risk assessment [5][6]. Nowadays, there are many technological advances, expanded Internet, GIS and relevant spatial information technologies along with great availability of open-source data and software. New methodologies and algorithms are able to process huge datasets and different types of information.

The modern GIS technologies allow instantaneous exchange of spatial information through Web-GIS platforms providing access to various research communities, experts, professionals and the wide public. GIS are able to present risk related information on various spatial and temporal scales. Such an integrated platform includes Decision Support System (DSS) to assist decision makers through interactive tools to better understand and handle risk management issues and at the same time to be the informative channels for local and regional agencies, as well as the wide public [7]-[13].

This study presents the design of a fully automated Web-based GIS platform, which is developed for river flood monitoring and risk prediction along the riversides of Arachthos (Figure 1). The short paper is organized as follows. Section II describes the study area and the main categories of data included in Web-based Geographical Information System for real-time flood monitoring. Section III presents the basic characteristics of the system and the expected final outcomes. We conclude the paper in Section IV.

II. DATA AND STUDY AREA

The Arachthos river is located in the northwestern Greece and it has the source in the mountain ridge of Pindus. This river is about 100-kilometers long and is the 8th biggest river in Greece. The river basin covers an area of 2.000 km^2. The climate of the greater area is characterized as typical Mediterranean with relatively mild and humid winter and very warm summer period. Arachthos delta is an area of great ecological interest and is protected by national / international legislation about ecosystems. Because of the geomorphology and the great water resources of Arachthos basin, two hydroelectric water dams have been constructed and used since 1981. The dams are located a few kilometers above the city of Arta, in the southern part of the river flow, which leads to the river delta in the Amvrakikos sea gulf. Although water dams could control the river flow, there are several flooding events. The analysis of river flow and estimation of flood risk is a complex process and highly dependent on the local topography. Thus, a careful collection of data with high-accuracy is needed along with a thorough pre-processing analysis. Table 1 presents the main datasets used to record the profile of the area and are covering all the needs for flood risk analysis.

Figure 1. The Epirus region (on the left). In its southeastern part, the Arachthos river is located. The river flows out in the Amvrakikos gulf (on the right).

The data is obtained from official national databases, governmental agencies and organizations and have been evaluated by our science team members. The datasets are either in vector or raster format. These datasets have been already checked regarding their quality and they have been used to produce additional layers of information so as to enrich their quality, through spatial editing procedures. Additionally, specific instrumentation to measure the river height and the meteorological conditions along the river will be installed and will operate in real-time basis in order to monitor in real time the potential risk levels regarding the possibility of flood events.

TABLE I. PARAMETERS USED TO DEVELOP THE MAIN GUI AS WELL AS FOR THE MULTICRITERIAL ANALYSIS OF RISK LEVELS PREDICTION.

a/a	Parameter	a/a	Parameter
1.	Road network	6.	Precipitation
2.	Urban fabric	7.	Topography (height, slope)
3.	Hydrological network	8.	Census data (population)
4.	Drainage basins	9.	Land use/land cover
5.	Geological information	10.	river sides

III. SYSTEM DESIGN

The Web-based Geographical Information System (Web-GIS) is composed of the main parts displayed in Figure 2. It is consisted of the following basic components: the client, the Web server, the functions and the database. The client allows the interaction of the Web-GIS users with the Web server and provides spatial information display. It comprises an Internet browsing software, which provides the service requesting and result visualization platform to the users on the client side. The server consists of a Web server and a database server.

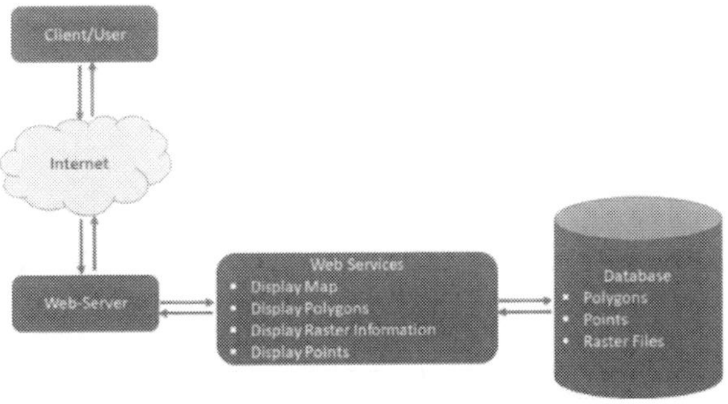

Figure 2. System architecture of the Web-GIS.

The Web server handles all the service requests and is responsible for transmitting the information back to the client. The Web server parses the HTTP requests from the clients and forwards them to the dedicated Web Service. The Web service query for the specific geospatial data from the database server and displays the data on the map. We use the Apache Web server, one of the most popular server development package and independent of the operating system (Apache, 2017). The scripting languages are PHP and Javascript. PHP (Hypertext Preprocessor) is a server-side scripting language designed for Web development and Javascript is a client-side scripting language, which creates interactive Webpages.

The database server is My-SQL (Structured Query Language), which is designed for high volume environments and it runs on all major operating systems. MySQL can contain both spatial and non-spatial data. It is able to handle high activity operations and provides support for geographic objects allowing location queries to be run in SQL [14].

The map display is based on Google Maps, which provides a geographically enabled base. In addition, the Google Map API [15] includes many built-in functionalities, which are used to create our own applications using the stored data. The viewer is composed of various data layer (functionalities), which will be included in the top frame, as well as in the side bar.

The flood risk layers of information will be provided at the advanced system through an automated multicriterial DSS, where real-time measurements will be integrated with the relative informational layer of Table 1, producing the final estimates. These flood risk layers will be automatically updated (dynamic layer of information) and will provide early-warning and decision support.

A few screenshots from the preliminary version of the Web-GIS platform are displayed below, Figures 3-5.

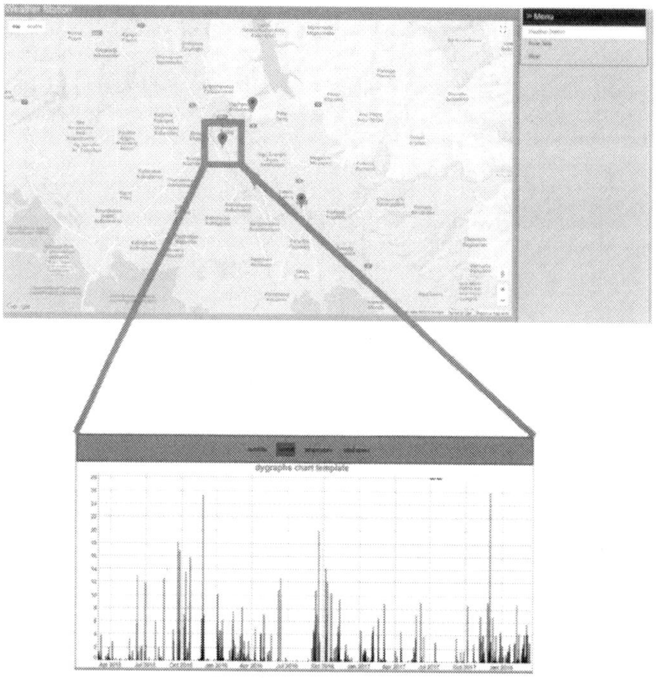

Figure 3. Display of Rainfall parameters of a metereological station based at Kampi, Arta.

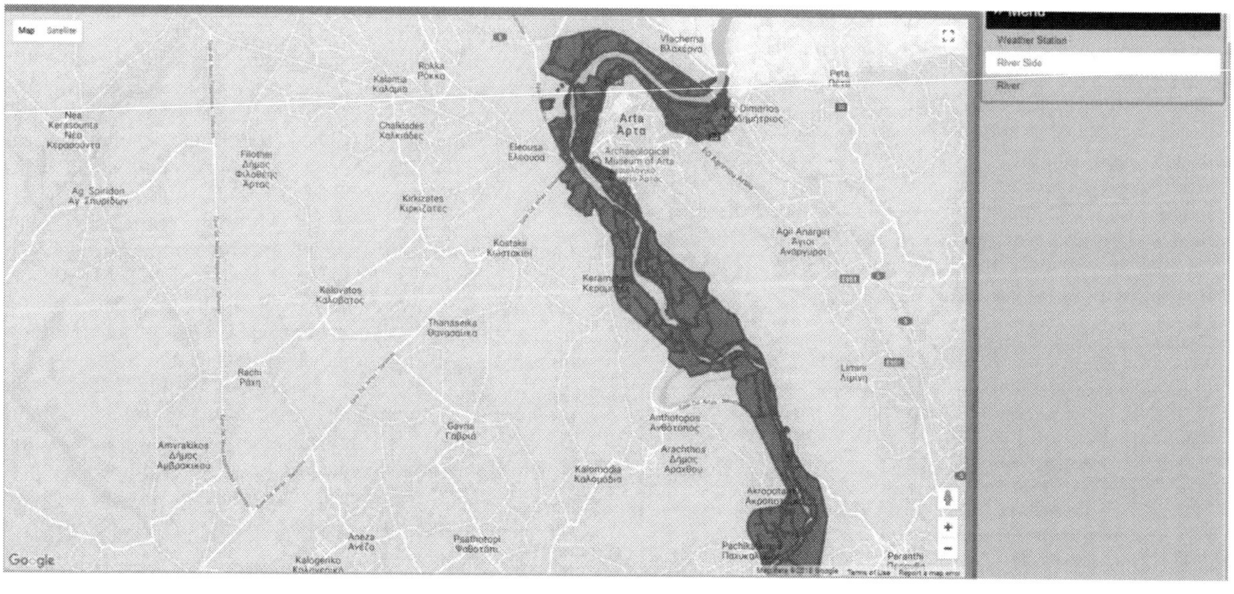

Figure 4. Display the River side of the river Arachtos, Arta.

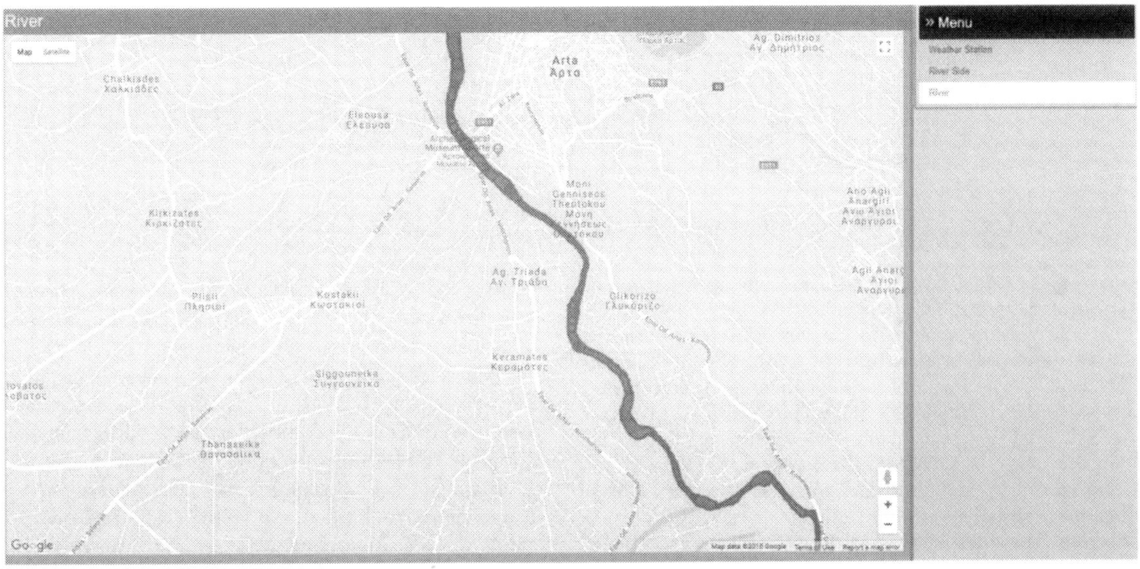

Figure 5. Display the River Arachtos, Arta.

IV. CONCLUSIONS

This study presents a Web-based GIS platform under development for monitoring the river flood and risk prediction along the riversides of Arachthos. Modern Information Communication Technologies (ICTs) are used to provide real-time information as well as early-warning risk levels regarding flood events.

ACKNOWLEDGMENTS

This study is co-funded by project "Integrated information system for flood events monitoring, management and early warning in the wider area of Arachthos" Epirus Operational Program 2014-2020, PA2 "Environmental Protection & Sustainable Growth", Co-financed by European Regional Development Fund (ERDF) and co-funded by project: "Enhancing research activities of Laboratory of Knowledge and Intelligent Computing" by Research Committee of TEI of Epirus.

REFERENCES

[1] K. Smith and R. Ward, "Floods: Physical Processes and Human Impacts", John Wiley and Sons, New York, 1998.

[2] EU, "Directive of the European Parliament and of the Council 2007/60/EC, establishing a framework for the assessment and management of flood risks", Official Journal 2007 L 288/27, European Commission, Brussels, 2007.

[3] J. Sunkpho and C. Ootamakon. "Real-time flood monitoring and warning system". Journal of Science & Technology, vol. 33, no. 2, 227-235, 2011.

[4] L. E. Colmenares-Guillen, O. N. Prieto and A. A. Jurado. "An Approach of Real-Time System for River Monitoring and Flood-Warning System in Puebla," Mexico. World Applied Programming, Vol. 3, Is. 8, August 2013, pp 328-340, 2013.

[5] S. Alaghmand, R. Abdullah, I. Abustan and B. Vosoogh. "GIS-based River Flood Hazard Mapping in Urban Area (A Case Study in Kayu Ara River Basin, Malaysia)". International Journal of Engineering and Technology vol. 2, no.6, pp 488-500, 2010.

[6] M. Abdeyazdan and A. Z. Jodaki. "Geographic Information System (GIS) Application in Flood Crisis Management". 2nd International conference on Innovative Engineering Technologies (ICIET'2015) August 7-8, 2015 Bangkok (Thailand), 2015.

[7] A. V. Frolov, V. V Asmus., S. V. Borshch, R. M. Vil'fand, I. I. Zhabina, V. V. Zatyagalova, V. A. Krovotyntsev, O. I. Kudryavtseva, E. A. Leont'eva, Y. A. Simonov and A. A. Stepanov. "GIS-Amur system of flood monitoring, forecasting, and early warning", vol. 41, Iss. 3, pp 157-169, 2016.

[8] Y. Xiao Y., S. Shanzhen and T. Zhongqian "GIS-based multi-criteria analysis method for flood risk assessment under urbanization," IEEE, 24th International Conference on GeoInformatics, 10.1109/GeoInformatics.2016.7578963, 2016.

[9] P. Fernandez, S. Mourato and M. Moreira, "Social vulnerability assessment of flood risk using GIS-based multicriteria decision analysis. A case study of Vila Nova de Gaia (Portugal)," Geomatics, Natural Hazards and Risk, vol. 7, no. 4, 1367-1389, 2016.

[10] A. Trono and L. Russo, Natural Disasters and sustainable development, Forecasts and use of new technologies to estimate natural disasters Authors: Edizioni Grifo.

[11] C. Stylios and P. Groumpos, Integrated Information System for Natural Disaster Management: Methodologies, Approaches, Good Practices. Patras Science Park, ISBN 978-960-85817-4-6, 2009.

[12] G. Georgoulas, P. Karvelis, S. Kolios and C. D. Stylios, Examining nominal and ordinal classifiers for forecasting wind speed. In: Proceedings of 8th IEEE Int. Conference on Intelligent Systems (IS'16), 3-5 September 2016, Sofia, Bulgaria. pp. 504-509, 2016.

[13] P. Karvelis, S. Kolios, G. Georgoulas and C. D. Stylios, Ensemble Learning for Forecasting Main Meteorological Parameters In: Proceedings of 2017 IEEE International Conference on Systems, Man and Cybernetics (SMC)), 5-8 October 2017, Banff Center, Banff, Canada, pp. 3711-3714, 2017.

[14] MySql, https://www.mysql.com/ [retrieved April 2017]

[15] Google MAP API, https://developers.google.com/maps/ [retrieved April 2017]

Software Quality Test Method of Satellite Control System using Software Test Automation System

Cheol Oh Jeong, Byoung-Sun Lee, In Jun Kim, Yoola Hwang, Soojeon Lee
Unmanned Vehicle Systems Research Group
Electronics and Telecommunications Research Institute (ETRI)
Daejeon, Rep. of Korea
E-mail: {cojeong, lbs, ijkim, ylhwang, soojeonlee}@etri.re.kr

Abstract—**Software (SW) quality management is a quality protection activity applied throughout SW development. SW quality management enables to implement software that is defect-free and meets development requirements. Quality management is performed through quality requirements, quality assurance methodology and quality testing. This paper shows SW quality test methods and results applied at the development stage of satellite control system. SW quality test was carried out by static test and dynamic test according to Verification-Validation (V-V) model proposed in SW engineering. The static test verifies that the source code is well implemented according to the requirements and design at the verification stage, and the dynamic test confirms in the validation phase whether the implemented product is running well. In order to efficiently perform the static test and the dynamic test, an SW test automation system is implemented. In this paper, we introduce a SW test automation system implemented for SW static test and dynamic test execution. Also, we show the results of the quality test that resulted in satisfying the requirements through potential errors or defects in the source code derived from static tests and requirements-based coverage results derived from dynamic tests.**

Keywords-Quality; Test; Satellite Control System (SCS); Static Test; Dynamic Test; Test Automation System; Coverage.

I. INTRODUCTION

The function and role of the software, which has been recognized as an auxiliary role of the hardware in the past, is gradually becoming a key factor that determines the completeness of the product. Accordingly, efforts are being made to ensure software quality from the development stage in order to secure software quality and efficient quality control. The satellite control system continuously monitors the state of the satellite during the operation of the satellite and periodically sends commands to the satellite to carry out a key function that enables the satellite to operate normally during the life of the satellite. Therefore, the SW of the satellite control system requires quality control from the development stage, so that the stability of the system should be secured. It is necessary to construct and operate a test automation system to efficiently perform the quality test of the SW requiring high reliability. The test automation system consists of a static test function that verifies the coding rule violation, code duplicate and complexity and a dynamic test function that checks the requirement based coverage required by the international standard [4]. Static testing is a technique for analyzing design documents and source code without direct execution of SW and examining whether coding rules, memory errors, duplicate code and code complexity are implemented according to international standards through coding standards, checklists, and so on [5]. On the other hand, the dynamic test is a technique to find SW defects by comparing the actual results with the expected results after executing the SW using test cases [6]. In this paper, we introduce a static test system and dynamic test system that verifies and validates the SW quality of the satellite control system. We present SW quality results through SW error and latent defect analysis derived from static test results, as well as coverage ratio analysis derived from dynamic test results.

The rest of the paper is structured as follows. In Section 2, we present an overview of SW quality test and SW test automation system including SW static test flow and dynamic test flow. In Section 3, we show the results of SW static test and dynamic test. In Section 4, we show the derived result of SW quality analysis from SW test results. Finally, in Section 5, we present conclusion and future work.

II. SW QUALITY TEST

A. *Overview of SW Quality Test*

SW quality control by V-V technique is performed to verify "Is the product being built correctly?" during verification phase and to validate "Is the right product being built?" during validation phase according to development and testing stages [1][3][4]. In the verification stage, the validity of the coding according to the specification is verified using the static test. In the validation stage, dynamic tests confirm whether the implemented SW satisfies the specification and requirements, as shown in Figure 1 [2]. Software test automation system was implemented and applied to verify and validate software quality efficiently.

B. *Overview of SW Test Automation System*

Software test automation system was implemented in Windows 7 environment, and it consists of Redmine (open source software) [8], Git (open source software) [9], Continuous Integration Automation Test (CIAT) and Web-based Quality Management System (WQMS), as shown in Figure 2. REDMINE is an issue management tool that manages issues that arise during both static and dynamic testing. GIT is a source configuration control tool that checks the version state of committed SW. CIAT is Continuous Integration (CI) server that performs static and dynamic tests and WQMS is a web-based quality monitoring system that allows administrators or development departments to constantly monitor static and dynamic test results and to control test quality and manage schedules [3].

Figure 3 shows the functional flow diagram of the SW test automation system. Test verification of the SW begins with the developer or the coding authority connecting directly to the test automation system and committing the SW. Once the SW is committed, the tests are run automatically after checking for the latest version. After the test is over, the test results will be distributed via e-mail to developers, administrators, managers and others involved in SW development. If there are errors present in the test result, the error will be corrected and the quality of the development SW is verified through retest.

SW test automation systems can perform static and dynamic tests. Table I shows test items which can be performed using the SW test automation system. For static tests, one can test for coding rules, memory errors, duplicate code, and code complexity. For dynamic tests, one can test for coverage. Tests can be performed using a test automation system for SWs implemented in C # languages [5][6].

TABLE I. TEST ITEMS OF SW TEST AUTOMATION SYSTEM

Test	Test Item	Tool
Static Test	Coding Rule Verification	FxCop
	Memory Error Verification	Sparrow QCE
	Duplicate Code Verification	CPD
	Code Complexity Verification	CCM
Dynamic Test	Coverage Measurement	SquishCoCo

The static test is performed according to the static test flow, as shown in Figure 4.

We derive dedicated SW coding rules for the static test by selecting the coding rules of error with high frequency of occurrence and the rules of high importance. The selection criteria for coding rules are as follows.

Rule selection criteria
- Rules that can be linked to faults when faults are dynamic
- Rules that may cause maintenance problems
- Rules that developers can easily understand

Rule exclusion criteria
- Rules that apply only to Visual Basic
- Rules that apply only to C / C ++
- Rules that difficult and abstract to explain
- Rules that proof is tough

Selected rules applying to static test are shown in Table II [7].

TABLE II. RULES SELECTION BY SELECTION CRITERIA

Coding Rule Verification

Tool	Rule Description	Applied Rules
FxCop	Design Rules	26
	Globalization Rules	1
	Interoperability Rules	11
	Maintainability Rules	0
	Mobility Rules	0
	Naming Rules	0
	Performance Rules	1
	Portability Rules	1
	Reliability Rules	0
	Security Rules	20
	Usage Rules	16
Total Applied Rules		76

Memory Error Verification

Tool	Rule Description	Applied Rules
Sparrow	API Usage	1
	Design	1
	Forbidden	1
	Misuse	3
	Program Crash	4
	Quality	24
Total Applied Rules		34

The requirement-based SW dynamic test was performed by the procedure shown in Figure 5.

First, the test items were derived from the requirements specification analysis for the system development, and then the test cases were designed to enable dynamic test input and output verification. A test case was implemented in a script format using a macro program so that the test automation system and the test case can be operated in an interlocking manner. The dynamic test system was performed by linking the implemented test case with the test automation system. The test cases implemented in the script format were implemented considering the test procedures and the efficient recursive test. Dynamic tests were performed on all the user interfaces of the implemented system. The test case implementation uses AutoHotKey, a macro program tool, and the scripted task case performs the dynamic test in the order of Setup, Start, Output, and End as shown in Figure 6.

The Setup phase copies the necessary Data Base (DB) and data files before starting the test, and the Start step compares the actual results with the expected results while performing the scenarios in the test procedure based on the user interface. The Output stage compares the actual results with the expected results for the entire procedure to determine the success of the test. Finally, the End step executes the test case script as a procedure to initialize the test.

III. RESULTS OF STATIC TEST AND DYNAMIC TEST

A. *Static Test*

Code rule verification and memory error verification were performed on a total of 80,602 lines of source codes according to the static test procedure. The static test execution result is provided through the expression screen of the test automation system to manager and SW developers, as shown in Figure 7.

Static test results are shown in Table III. A total of 871 rule violations were detected in the coding rule verification including all violations from critical warning, critical error, and error violation. A total of 3,340 rule violations were detected in memory error verification, including all violations from level 1 to level 3.

TABLE III. STATIC TEST RESULTS

Coding Rule Verification Results

Coding Rule	# of Detection
Design Rules	768
Globalization Rules	0
Interoperability Rules	0
Performance Rules	3
Portability Rules	0
Security Rules	0
Usage Rules	100
Total Violation Number	871

Memory Error Verification Results

Memory Error Rule	# of Detection
API Usage	0
Design	0
Forbidden	0
Misuse	0
Program Crash	170
Quality	3170
Total Violation Number	3,340

B. *Dynamic Test*

Dynamic testing was performed on a requirement based basis and was performed on a flight dynamics subsystem. A total of 20 test cases were derived for the 10 test items through the requirements analysis for the dynamic test execution. The errors/faults that are found during the dynamic test are delivered to the developer and the recursive test is performed to confirm the normal operation of the system through the retest after the error correction. The result of the dynamic test is provided through the display screen of the test automation system, as shown in Figure 8.

Dynamic test results of performing a requirement based dynamic test are shown in Table IV. Of the total of 20 test cases, 18 test cases have been confirmed to pass successfully.

TABLE IV. DYNAMIC TEST RESULTS

Test Case ID	Test Description	Pass/Fail
TFDS-01	Six-hour Orbit Prediction without Maneuver	Pass
TFDS -02	One-month Orbit Prediction without Maneuver	Pass
TFDS -03	One-month Orbit Prediction with Maneuver	Pass
TFDS -04	Two Stations Bias Calibration	Pass
TFDS -05	Orbit Determination Using Commercial Data without Maneuver	Pass
TFDS -06	Orbit Determination including Maneuver	Pass
TFDS -07	Real-time Orbit Determination	Pass
TFDS -08	Event Prediction (Eclipse & Sun-interference test)	Pass
TFDS -09	Station-Keeping Maneuver Planning and Reconstruction	Pass
TFDS -10	Station-Keeping Maneuver Preparation	Pass
TFDS -11	Station-Relocation Maneuver Planning	Pass
TFDS -12	De-Orbit Maneuver Planning	Pass
TFDS -13	Collision Avoidance	Pass
TFDS -15	Fuel Accounting by TOT Method	Pass
TFDS -17	H Management	Pass
TFDS -18	On-board Orbit Propagator	Pass
TFDS -19	Earth Acquisition	Pass
TFDS -20	Plot Services	Pass
TFDS -23	Data Management	Fail
TFDS -24	Process Management	Fail

IV. SW QUALITY ANALYSIS RESULT

The SW quality level of static test can be confirmed by the following formula.

Violation ratio (%) = # of violations/total line x100

When the SW quality level of static test is derived by reflecting the number of violations in Table III for a total of 80,602 lines of source code, the violation ratio of about 1% has calculated for the coding rule verification and about 4% for the memory error verification has calculated, as shown in Table V.

TABLE V. SW QUALITY LEVEL OF STATIC TEST

Classification	# of Source Code Line	# of Violation	Violation Ratio
Coding Rule Verification	80,602	871	1.08%
Memory Error Verification		3,340	4.15%

The SW quality of static test results can be considered to be high. It is possible to implement a more reliable SW by debugging the violation code with the recommended code format suggested by the coding rule format.

The SW quality level of dynamic test can be confirmed by the following formula.

Coverage ratio (%)=# of success cases/# of total test cases

When the SW quality level of dynamic test is derived by reflecting the number of success cases of 18 for a total test cases of 20, 90% of coverage ratio is calculated. The dynamic test was terminated when reaching a preset coverage target of 90%. The quality of the dynamic test results is considered to be high as the preset coverage target is achieved. As the preset coverage target was achieved, two test cases were not performed dynamic test and were marked as fail. By re-configuring coverage targets and performing source code debugging and dynamic testing, we expect 100% coverage to be achieved.

V. CONCLUSION & FUTURE WORK

This paper introduced the SW quality test method of SCS using SW test automation system implemented to ensure the quality and stability of software during the software development period. We also presented the static test results and dynamic test results as well as SW quality analysis results. This allows for the assurance of quality and stability of the development software which was complemented by identifying and modifying the software errors and defects during the development period. Through this, it is possible to analyze SW quality of satellite control system, and SW quality and stability are secured. When participating in similar satellite control system development program in the future, it is expected to apply the implemented software test automation system to ensure that the quality and reliability of mature software development is applicable.

ACKNOWLEDGMENT

This work was supported by the Space Core Technology Development Program of NRF-Ministry of Science, ICT and Future Planning [NRF-2014M1A3A3A03034729, Development of core S/W standard platform for GEO satellite ground control system]

REFERENCES

[1] National IT Industry Promotion Agency. nipa: SW Development Quality Management Manual. [Online] Available from: http://codedragon.tistory.com/attachment/cfile22.uf@233B814954902830161BD7.pdf, Mar. 17, 2018

[2] Electromics & Telecommunications Research Institute, ETRI: Technical Document - SW Quality Management System of SGC Core Platform Development, Jan. 2015

[3] C. O. Jeong, B. S. Lee, I. J. Kim, Y. L. Hwang, and S. J. Lee, "Automated S/W testing system for Core S/W of Ground Control System", KOSST Conference, Jun. 2015

[4] C. O. Jeong, B. S. Lee, I. J. Kim, Y. L. Hwang, and S. J. Lee, "SW Quality Test Method of Satellite Control System", The Korean Society for Aeronautical and Space Sciences, Fall conference, Nov. 2015,

[5] C. O. Jeong, B. S. Lee, I. J. Kim, Y .L. Hwang, and S. J. Lee, "SW Static Test Method of Satellite Ground Control System", The Korean Society for Aeronautical and Space Sciences, Spring conference, Apr. 2016

[6] C. O. Jeong, B. S. Lee, I. J. Kim, Y. L. Hwang, and S. J. Lee, "Software Dynamic Test Environment Development of Satellite Ground Control System", The Korean Society for Aeronautical and Space Sciences, Fall conference, Nov. 2016

[7] C. O. Jeong, B. S. Lee, I. J. Kim, Y. L. Hwang, and S. J. Lee, "SW test automation system implementation for securing SW quality and stability", Int'l Conference on Software Engineering Research and Practice (SERP'17), Jul. 2017

[8] REDMINE: [Online] Available from: http://www.redmine.org/projects/redmine/wiki/Download, Mar. 17, 2018

[9] Git: [Online] Available https://git-scm.com/downloads, Mar. 17, 2018

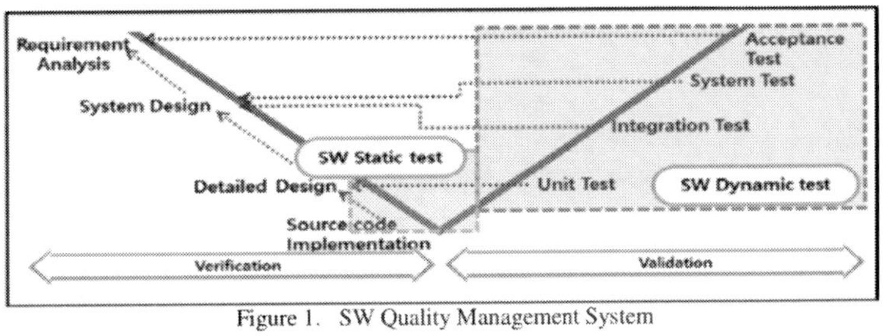

Figure 1. SW Quality Management System

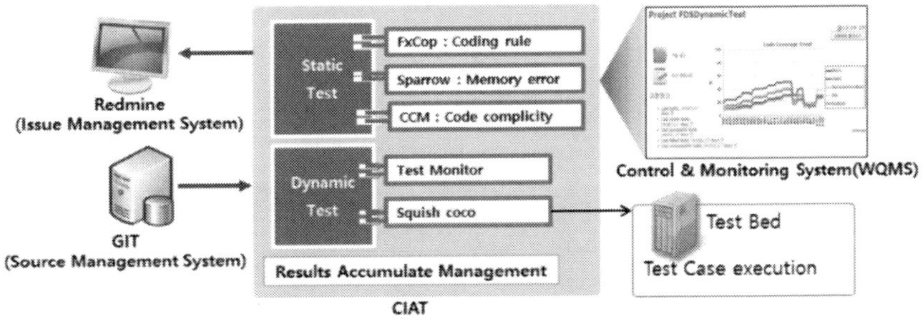

Figure 2. Test automation system configuration

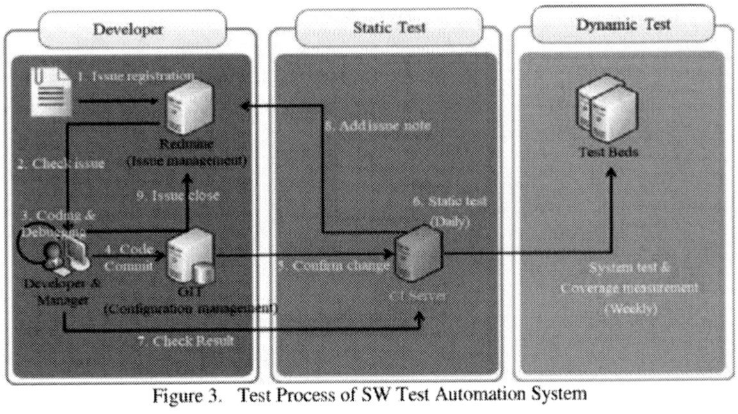

Figure 3. Test Process of SW Test Automation System

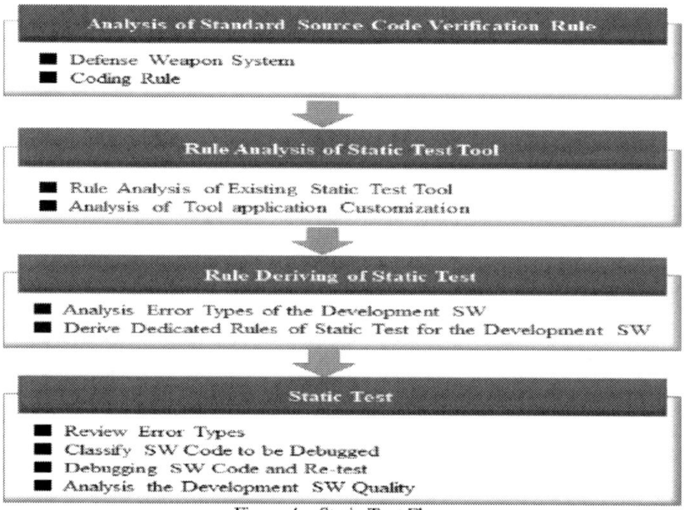

Figure 4. Static Test Flow

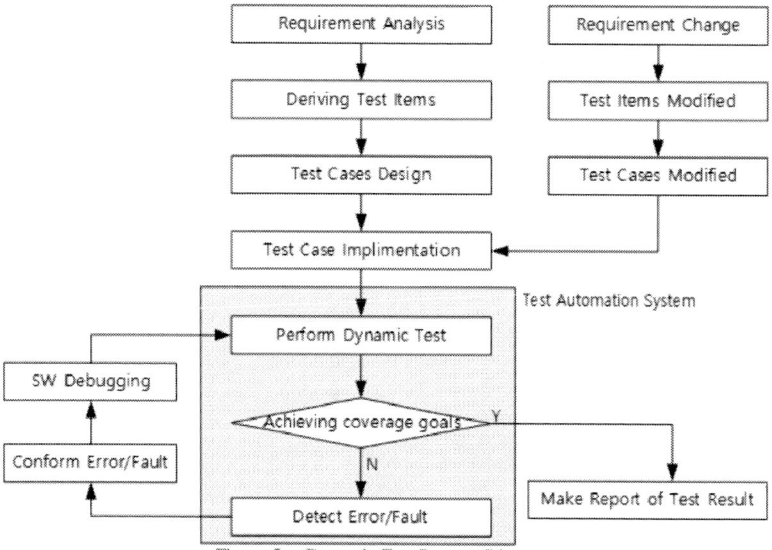

Figure 5. Dynamic Test Process Diagram

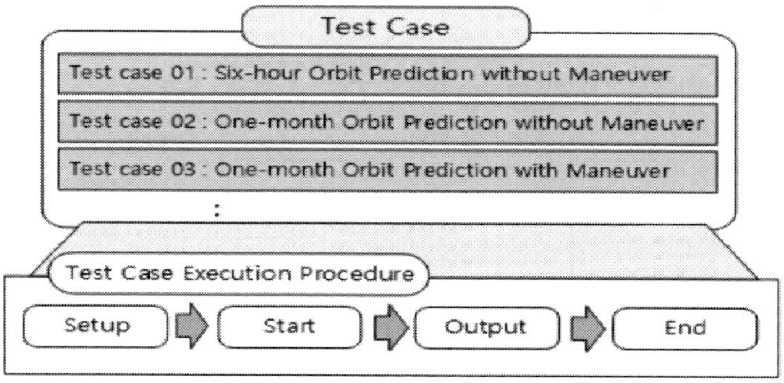

Figure 6. Test Case Script Structure

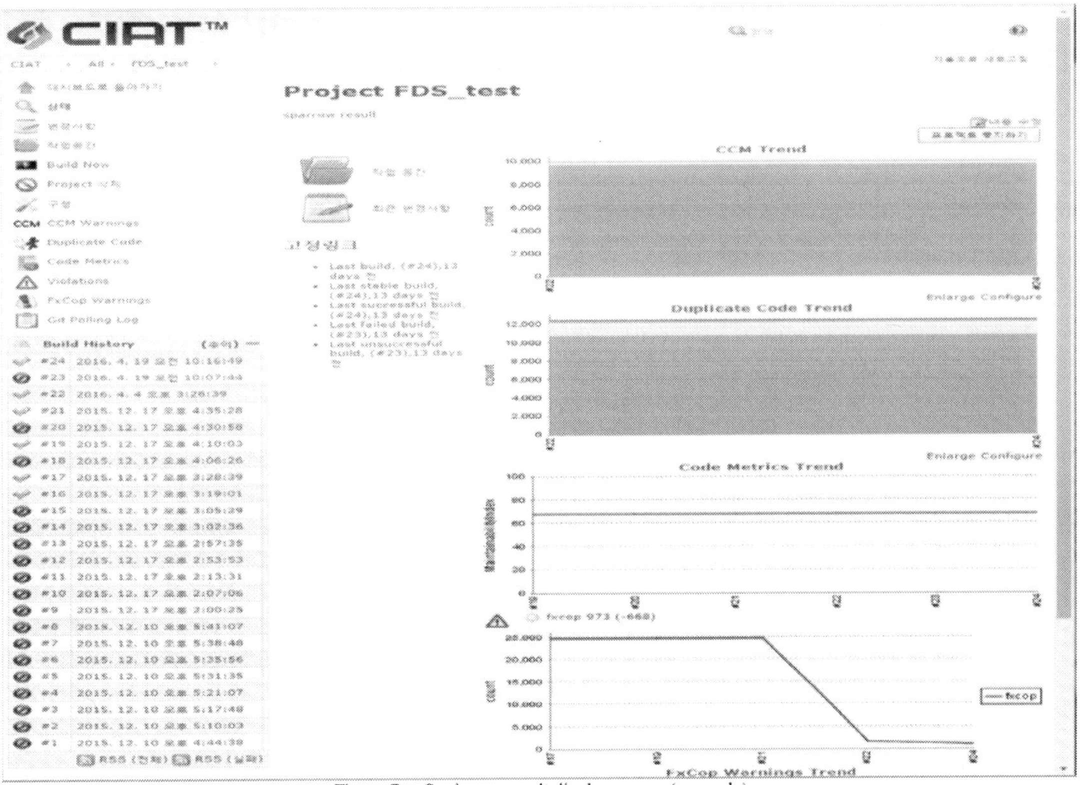

Figure 7. Static test result display screen (example)

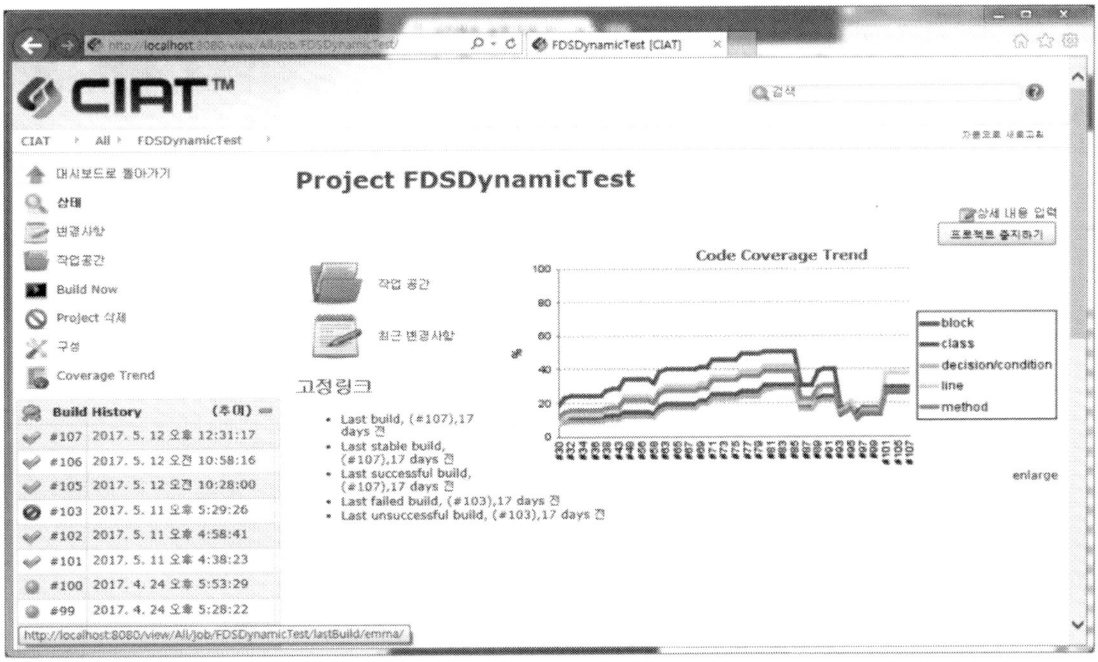

Figure 8. Dynamic test result display screen (Example)

The Effect of Radar Cross Section and Speed of Target on the Detection of MIMO Radar

Raed S. M. Daraghma
Department of Electrical and Electronics Engineering
Palestine Technical University, Palestine
Email: R.daraghmeh@ptuk.edu.ps

Abstract— **In this paper, we address the problem of Direction of Departure (DOD) and Direction of Arrival (DOA) estimation for Multiple Input Multiple Output (MIMO) radar. The work presented studies the effect of Radar Cross Section (RCS), Signal to Noise Ratio (SNR) and speed of targets on the performance of the MIMO radar. Analysis can be used to find the direction of multiple types of targets such as CAPON, MUSIC and parallel factor (PARAFAC). To differentiate the meaning of targets, varying targets of different types, such as bicycle, bird, man, ship and jet have been considered. After defining suitable values for each type of target in 2D space, the performance of each type is discussed by using the MATLAB program.**

Keywords- MIMO Radar; Target Localization; Parallel Factor (PARAFC); Direction of Arrival.

I. INTRODUCTION

Some of the most important RADAR applications are the detection performance and high resolution of the moving target localization. Radar Cross Section (RCS), range, location and velocity are utility parameters of the moving target [1][2]. To improve the accuracy of target detection and estimation, antenna arrays have been used. MIMO radar uses multiple transmitter and multiple receiver elements. Generally, unlike the phased-array systems, MIMO radar has several advantages compared to the conventional phased array systems: higher resolution, more degrees of freedom, improved parameter specification, better spatial coverage and detection diversity gain. MIMO radars can be classified into two categories: (1) MIMO radar with widely separated antennas scheme and (2) MIMO radar using collocated antennas, which is similar to phase array radar [3][4]. In the literature, there are many configurations of MIMO radar according to the location of the transmitting and receiving elements. Widely separated antennas represent one of these configurations. In this scheme, the separation between transmitter and receiver should be large enough to receive the uncorrelated echoes from the different targets. The main advantage of this scheme is that the spatial diversity of the targets RCS enhances the radar performance [5]. In this paper, a bistatic MIMO radar technique with transmission spatial diversity is proposed, and the estimation performance is analyzed. Moreover, the angles with respect to receiver can be determined using the proposed technique. In addition, the maximum number of targets that can be identified with this technique is discussed in this paper. MIMO radar can deal with multiple targets. Linearly independent waveforms are transmitted at the same time via multiple antennas. These independent waveforms are linearly combined at the targets with different phases, after which the signal waveforms reflected from different targets are linearly independent of each other, which allow for the application of CAPON, MUSIC and PARAFAC algorithms [6]. In this work, we focus on the application of MIMO radar to the estimation of DOA and the DOD of multiple targets exist in the same range bin for bistatic MIMO radar system. We are particularly interested to optimize the average angular error for different types of targets.

The paper is organized as follows. In Section 2, the previous work on the subject is summarized. The MIMO radar signal model is presented in Sections 3 and 4. The performance of MIMO radar is evaluated through simulations via MATLAB in Section 5. Some concluding remarks are given in Section 6.

II. RELATED WORK

There are many existing methods to localize the moving target. The following parameters: the Angle of Arrival (AOA), the Angle of Departure (AOD), the speed of the target and the RCS are the most used to localize a moving target. Many techniques have been proposed such as CAPON technique [6], the MUSIC (Multiple Signal Classification) technique [7], and parallel factor analysis (PARFAC) [8].

In this paper, we select a number of different moving targets: simple and complex targets with different RCS and speeds. From the existing work on the application of CAPON, MUSIC and PARAFAC to the localization of different targets, we notice the importance of the types of targets and the effect of changing the speed of targets. This paper focuses on comparing the performance criterion for different types of targets as well as the impact of the number of antennas on the performance of three different techniques mentioned above.

III. MIMO RADAR SIGNAL MODEL

In this section, we consider that the Coherent Processing Interval (CPI) consists of Q consecutive pulse periods. The Swelling II target model [8] was assumed, where RCS coefficient is varying from pulse to pulse. The targets are located in the far-field. The RCS coefficients are assumed to vary independently from pulse to pulse, and the propagation medium is non dispersive.

The baseband received signal at the output of the receive array after synchronization can be written as:

$$X_q = B(\phi) \Sigma_q A^T(\Theta)S + W_q \text{ , where q = 1,.....,Q} \quad (1)$$

where $X_q \in \mathbb{C}^{M_r \times L}$ collects the L samples received by M_r antennas for the qth pulse period.

$\Sigma_q = diag\ (c_q)$, with c_q=[$\delta_{1q},...,\delta_{kq}$] where $\delta_{kq} = \alpha_{qk} e^{j(q-1)x_q}$, i.e., x_q is the Doppler frequency of the kth target [9]. The RCS coefficients α_{qk} , k = 1,...,k, are varying independently from pulse to pulse, and $W_q \in \mathbb{C}^{M_r \times L}$ is the noise interference term. MIMO radar transmits mutually orthogonal waveforms. We assume that $1/SLS^H = I_M$. After right multiplication of (1) by $\frac{1}{L}S^H$, the matched filter output is:

$$Y_q = B(\phi) \Sigma_q A^T(\Theta) + Z_q \text{ , where q = 1,.....,Q} \quad (2)$$

Where

$$Y_q = {}^1/_{LXS^H} \in \mathbb{C}^{M_r \times M_t} \text{ and } \quad Z_q = {}^1/_{LWS^H} \quad (3)$$

Let us factorize (3):

$$Y_q = (A(\phi) \odot B(\Theta) c_q^T) + Z_q \text{ where, q = 1,..., Q} \quad (4)$$

where

$$Y_q = vec\ (Y_q)\ ,\ Z_q = vec\ (Z_q) \quad (5)$$

which can be written in the compact form:

$$Y = (A(\phi) \odot B(\Theta)) C^T + Z \quad (6)$$

where Y=[$Y_1,...,Y_Q$] and Z= [$Z_1,...,Z_Q$] are of size $M_t M_r \times$ Q and C^T= [$C_1^T, ... C_Q^T$] is of size K × Q.

From [10] the CAPON estimator can be written as:

$$P(\phi,\Theta) = \frac{1}{(a(\phi) \odot b(\Theta)^H R_{YY}^{-1} (a(\phi) \odot b(\Theta))} \quad (7)$$

$$R_{YY} = \frac{1}{Q} YY^H$$

The MUSIC estimator can be written as:

$$P_{MUSIC}(\phi,\Theta) = \frac{1}{a(\phi) \odot b(\Theta))^H E_Y E_Y^H\ a(\phi) \odot b(\Theta))} \quad (8)$$

where E_Y is the $M_t M_r \times (M_t M_r - K)$ matrix contains the noise eigenvectors of R_{YY} .

The third estimator PARAFAC was derived in [11]. PARAFAC implies the transmit and receive angles relative to the same target are automatically paired.

IV. DATA MODEL

In this section, we consider the multiple pulses, multiple arrays case.The MIMO radar system has the following parameters :

- M_t transmit array.
- M_r receive array.
- K targets in a far field.
- Q transmitted pulses, and the RCS is varying independently from pulse to pulse (Swelling II case)
- δ_{kq} is the reflection coefficient of the Kth target during the qth pulse.
- $\{\Theta\}_{k=1}^K$, $\{\phi\}_{k=1}^K$ are the DODs and DOAs with respect to transmit and receive array, respectively.

A(Θ) = [a(Θ_1),...., a(Θ_k)] is the transmit steering vector relative to K targets, B(ϕ) = [b(ϕ_1),...., b(ϕ_k)] is receiving steering vector relative to K targets.

V. SIMULATION RESULTS

In this section, MATLAB program simulation results are presented to verify the above analysis and compare the performance of the three techniques (Capon, MUSIC and PARAFAC). Localization of the multiple targets for a Uniform Linear Array (ULA) configuration at the transmitter and receiver can be achieved by the above algorithms [8].

We generate the matrices S, A and B as explained in the previous section. S is generated by $[S]_m$ = (1+j/2) $[H_N]_m$, where H_N is the N × N Hadamard matrix, and N is fixed to 256.

The Signal To Noise Ratio (SNR) at the receiver is defined as: SNR = $10\log\left(\sum_{q=1}^{Q} \|B \Sigma_q A^T S\|^2 F / \|W\|^2 F\right)$ dB, where Additive White Gaussian Noise (AWGN) is assumed, and $\|.\|_F$ is the Frobenis norm. We consider ULA transmit and receive arrays with $\lambda/2$ interelement spacing for both arrays[8]. For the Swerling II target model, each column of the matrix $C \in \mathbb{C}^{Q \times K}$ is generated from a complex Gaussian distribution with zero mean and variance $\sigma^2_{\delta k}$,where α_k is sample drawn from a complex Gaussian distribution with zero mean and variance $\sigma^2_{\delta k}$ and the Doppler frequency x_k is generated by:

$$x_k = \frac{2\,\Pi V_k\, T_p}{\lambda}$$

where v_k is the target velocity, $T_p = 5\times10^{-6}$ is the pulse duration in seconds, $\lambda = 3 \times 10^8 / f_c$, with $f_c = 1$ GHz.

This subsection analyzes the impact of the number of targets on the performance detection. The performance criterion is the absolute value of the difference between the true angle and estimated angle, averaged over transmit and receive angles and over all targets. In a first experiment (Figure 1), we consider seven types of targets. The variance and speed of each target was given in Table I. We simulate the presence of two to six targets , starting from K=2 with DODs [10^0 , 20^0] and DOAs [0^0,30^0] until K=6, DODs [10^0 , 20^0 , 30^0 , 40^0 , 50^0 , 60^0] and DOAs [0^0, 30^0, 5^0, 15^0, 25 ,30]. The number of pulses is Q = 100, number of samples for each transmitted pulses is L =512, SNR =10 dB, and Swerling II model is chosen. We plotted the performance of the CAPON method, and we compared the performance of the different types of targets via Monte Carlo simulation. From Figure 1, it is clear that a better angular resolution is achieved when the target is "Man" and the worst angular resolution is achieved when the target is "Car". From Figure 1, we observe that the global performance of all types of targets seriously degrade when the number of targets is increased. In Figure 2, we simulate the presence of two to six targets. The other parameters are the same as in Figure 1, but, in this case we have plotted the performance of the MUSIC technique. We compare the performance of the different types of targets via Monte Carlo simulation. From Figure 2, it is clear that a better angular resolution is achieved when the target is "Boat" and the worst angular resolution is achieved when the target is "Fighter". From Figure 2, we observe that the global performance of all types of targets seriously degrades when the number of targets is increased.

In Figure 3, we simulate the presence of two to six targets. The other parameters are the same as in Figure 1, but, in this case, we have plotted the performance of the parallel factor (PARAFC) technique, and we compared the performance of the different types of targets via Monte Carlo simulation. From Figure 3, it is clear that the best angular resolution is achieved when the target is "Car" and the worst angular resolution is achieved when the target is "Jet". From Figure 3, we observe that the global performance of all types of targets seriously degrades when the number of targets is increased.

This subsection analyzes the impact of signal to noise ratio on the performance detection. In a second experiment (see Figure 4), we simulate the presence of three targets(K=3) characterized by DODs [10^0 , 20^0 ,30^0] and DOAs [-10^0,- 20^0 ,-30^0]. The number of pulses Q = 100, the number of samples for each transmitted pulse L =512, the number of transmit and receive sub arrays is fixed to 5, SNR ϵ (0,2,4,6,8,10) dB, and the Swerling II model is chosen. We plotted the performance of the CAPON method, and we compared the performance of the different types of targets via Monte Carlo simulation. From Figure 4, it is clear that the best angular resolution is achieved when the target is "Bird" and the worst angular resolution is achieved when the target is "Bicycle". As expected, from Figure 4, we observe that the performance of all types of targets improves when the signal to noise ratio increases. In Figure 5, we simulate the presence of three targets. The other parameters are the same as in Figure 4, but, in this case, we have plotted the performance of the MUSIC technique, and we compared the performance of the different types of targets via Monte Carlo simulation. From Figure 5, it is clear that a better angular resolution is achieved when the target is "Bird" and the worst angular resolution is achieved when the target is "Bicycle". In Figure 6, we simulate the presence of three targets. The other parameters are the same as in Figure 4, but, in this case, we have plotted the performance of the parallel factor (PARAFAC) technique and we compared the performance of the different types of targets via Monte Carlo simulation. From Figure 6, it is clear that the best angular resolution is achieved when the target is "Bird" and the worst angular resolution is achieved when the target is "Jet".

TABLE I. RCS AND SPEED FOR DIFFERENT TYPES OF TARGETS

Target Type	Radar cross section fot target (in m^2)	Speed of target (m/s^2)
Bicycle	2	10
Man	1	6.5
Car	100	100
Fighter	40	125
Boat	0.02	20
Jumbo Jet	100	40
Bird	0.01	150

VI. CONCLUSION

In this paper, we have considered the detection and localization of moving target in bistatic MIMO radar with widely separated antennas, where multiple antennas transmit linearly independent waveforms and multiple antenna receive the reflected signal. We can significantly improve the estimation accuracy of the bistatic MIMO radar techniques as well as enhance their performance. The main problems encountered in MIMO radar detection are radar cross section and speed of the target. To illustrate the impact of these two parameters on the performance of MIMO radar, several types of targets and three popular techniques (CAPON, MUSIC and PARAFC) were considered for comparison. From the simulation results, we have shown that irrespective of the radar cross section and speed of target a high performance (low angular error) can be obtained when the signal to noise ratio increases. On the contrary, low performance can be obtained when the number of targets increases.

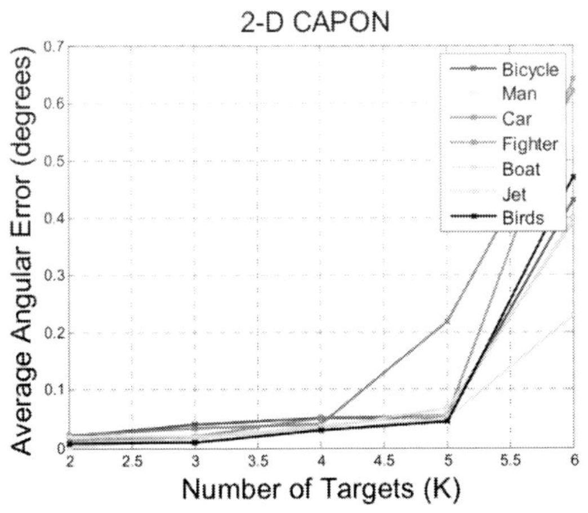

Figure 1. Average angular error with number of targets (2-D capon case)

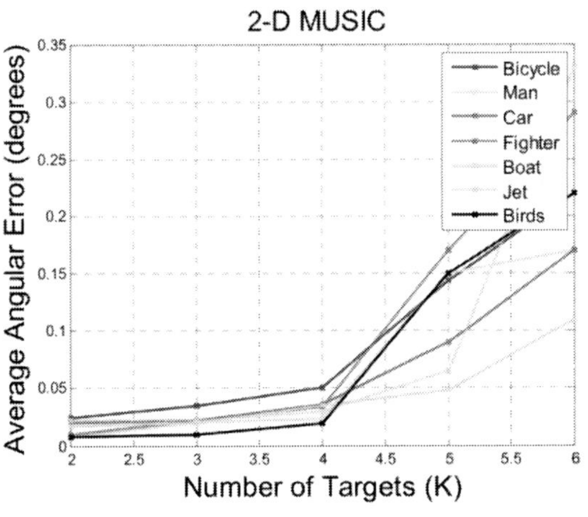

Figure 2 . Average angular error with number of targets (2-D capon case)

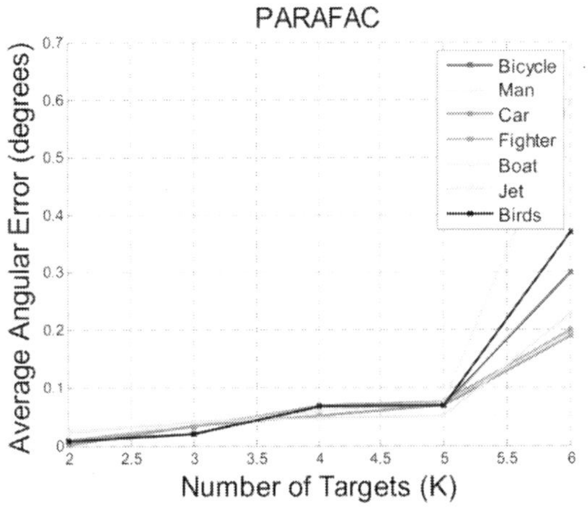

Figure 3. Average angular error with number of targets (PARAFAC case)

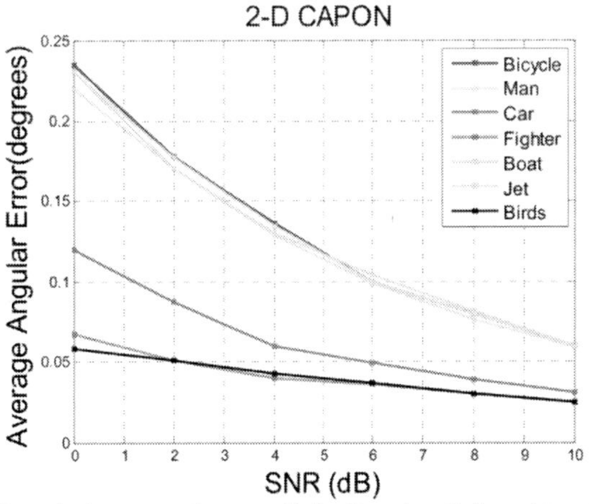

Figure 4. Average angular error with signal to noise ratio for each target (2-D Capon case)

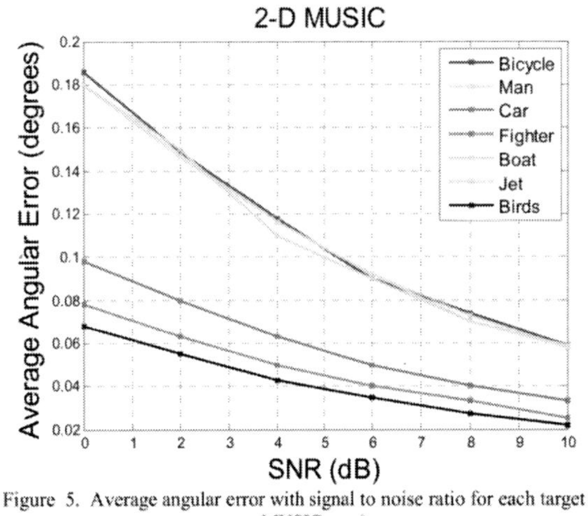

Figure 5. Average angular error with signal to noise ratio for each target (2-D MUSIC case)

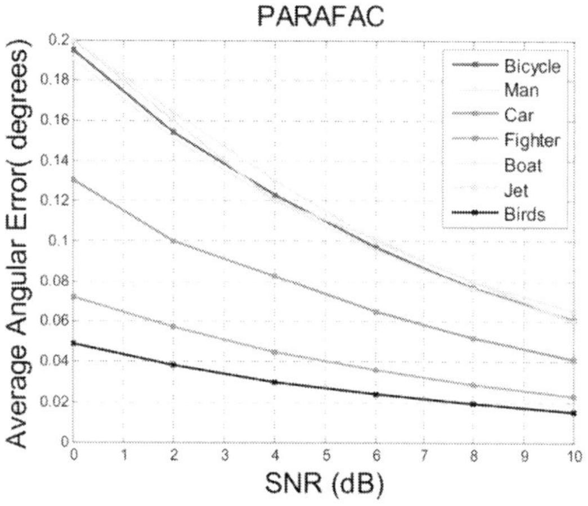

Figure 6. Average angular error with signal to noise ratio for each target (PARAFAC case)

REFERENCES

[1] G. R. Curry, Radar System Performance Modeling, 2nd ed. Norwood, MA; Artech House, 2005.

[2] M. I. Skolnik, Introduction to Radar Systems, 3rd ed. New York; McGraw-Hill, 2001.

[3] D. W. Bliss and K. W. Forsythe, " Multiple-input multiple-output (MIMO) radar and imaging: degrees of freedom and resolution", in Proc.37th Asilomar Conference. Signals, Systems., computer., Pacific Grove, CA, Nov. 2003, vol. 1, pp.4-59.

[4] F. C. R obey, S. Coutts, D. Weikle, J. C. McHarg, and K. Cuomo, " MIMO radar the theory and experimental results", in Proc. 38th 37th Asilomar Conference. Signals, Systems., computer., Pacific Grove, CA, Nov. 2004, vol. 1, pp.300-304.

[5] E. Fishler et al. , " MIMO radar: an idea whose time has come, " in Proceedings of the IEEE Radar Conference, PP. 71-78, Philadelphia, Pa, USA, April 2004.

[6] J. Capo, " High-resolution frequency- wave number spectrum analysis". Proceedings of the IEEE, 57(8): 1408 { 1418, 1969. doi:10.1109/PROC. 1969.7278}.

[7] R. O. Schmidt, "Multiple emitter location and signal parameter estimation". IEEE Transactions on Antennas and Propagation,34(3): 276 { 280, 1986. ISSN 0018-926X. doi: 10.1109/TAP. 1986.1143830} .

[8] G. Cichocki et al. "Tensor decompositions for signal Process"., philadelphia, pa USA april. 2015, 32, 145-163.

[9] C. Y. Chen and P.P. Vaidyanathan, " MIMO radar space-time adaptive processing using prolate spheroid wave function", IEEE trans. Signal Process., vol. 56, no. 2, pp. 623-635, Feb. 2008.

[10] H. Yan, J. Li, and G. Liao, " Multitarget identification and localization using bistatic MIMO radar systems", EURASIP L. Adv. In Signal Process. pp 77-78, vol. 2008, no. ID 283483, 2008.

[11] D. Nion and N. D. Sidiropoulos, " A PARAFAC- based technique for detection and localization of multiple targets in a MIMO radar system," in Proc. IEEE Int. Conf. Acoust., Speech Signal Process. (ICASSP).pp2077-2080, 2009.

GPS AOA Estimation Technique Based on a Null Despreader

Suk-seung Hwang
Department of Electronic Engineering,
Chosun University,
Gwangju 501-759, Republic of Korea
email: hwangss@chosun.ac.kr

Abstract—**The Global Positioning System (GPS), which has various military and civilian applications, is designed to estimate the locations of objects including the user. In order to enhance the desired GPS signal and to efficiently suppress the interference signals, the Angle of Arrival (AOA) information for the GPS signal is required on the environment of high-power interferers. After despreading based on a conventional despreader, the GPS AOA can be estimated using the conventional AOA estimation algorithm such as Multiple Signal Classification (MUSIC). However, we must select the GPS AOA from all estimated AOAs, because the result of MUSIC includes AOAs of GPS and interference signals. In order to overcome this problem, in this paper, we propose the GPS AOA estimation algorithm based on a null despreader. In this technique, we compare the estimated AOAs based on the conventional despreader and based on the null despreader. The proposed technique efficiently estimates and selects the desired GPS AOA when existing high-power interference signals. We demonstrate the performance of the proposed technique via a representative computer simulation example.**

***Keywords-Global Positioning System; Angle-of-Arrival; null despreader; Multiple Signal Classification; beamforming*.**

I. INTRODUCTION

The Location Detection Technology (LDT) is utilized in a variety of industrial areas and GPS is one of the representative LDTs for estimating the location of users or objects [1]-[3]. Since GPS employs the low-power Direct-Sequence Spread-Spectrum (DSSS) signal, its Signal-to-Noise Ratio (SNR) is very low (typically as low as -30 dB) and it suffers from the high-power jamming (interference) signals, which Jammer-to-Signal Ratio (JSR) can exceed 40 dB [4]. Also, interference signals have various different forms, such as Continuous Waveform (CW) and Frequency Modulated (FM) signals as the Constant Modulus (CM) signal, a Wide-Band (WB) noise jammer, and a pulsed jammer which has periodic on/off characteristics [5][6]. Although the low SNR problem can be solved by the large spreading gain (≈ 43 dB) of GPS, the high JSR problem might cause the serious location estimation error for the conventional GPS receiver. In order to solve this problem, we employ the adaptive beamforming techniques based on the GPS AOA information, such as the Minimum-Variance-Distortionless-Response (MVDR) [7][8] and the adaptive Generalized Sidelobe Canceler (GSC) [9][10].

Since the GPS signal power before despreading is lower than the noise power level, it is not possible to estimate the GPS AOA using the conventional AOA estimator such as MUSIC. After despreading, the GPS AOA can be estimated using the MUSIC algorithm because it is higher than the noise level, but we must select the GPS AOA from all estimated results because they include AOAs of the high-power interference signals. In this paper, we propose the GPS AOA estimation technique based on a null depsreader, which efficiently estimates and selects the GPS AOA information in the high JSR environment. The estimated results based on the null despreader do not include the GPS AOA but include AOAs of interference signals, because it is designed to reject the GPS signal while retaining the other interference signals [11]. On the other hand, the estimated results based on the conventional despreader include AOAs of the GPS signal and interferers. In order to select the GPS AOA, we compare both results based on the conventional and null despreaders, and determine the AOA included in results of the conventional despreader but excluded in results of the null despreader, as the GPS AOA.

The rest of this paper is organized as follows: In section II, we define the received signal model for the GPS signal, interference or jamming signal, and Additive White Gaussian Noise (AWGN). Section III describes the proposed GPS AOA estimation algorithm based on the null despreader. The performance of the proposed technique is discussed by the representative computer simulation results in Section IV. Finally, conclusions are presented in Section V.

II. RECEIVED SIGNAL MODEL

In this paper, we employ an antenna array with M elements at the receiver and focus on estimating the AOA of the GPS Coarse Acquisition (C/A) code. At discrete sample index k, the received signal vector can be modeled as

$$\mathbf{r}(k)=\mathbf{d}_c c_i(k) b_i(k)+\mathbf{D}\mathbf{s}(k)+\mathbf{v}(k), \qquad (1)$$

where $c_i(k)$ is an element of the cyclostationary pseudorandom noise (PRN) code (length $N = 20\times1023$), $b_i(k)$ is the GPS data bit which remains constant over the length of one cycle of the PRN code, for the i th satellite, and $\mathbf{v}(k)$ is the AWGN vector (size M) with independent and identically distributed components, each with zero mean and variance σ^2. The other quantities and sizes of matrices and vectors in (1) are summarized in Table I, where the

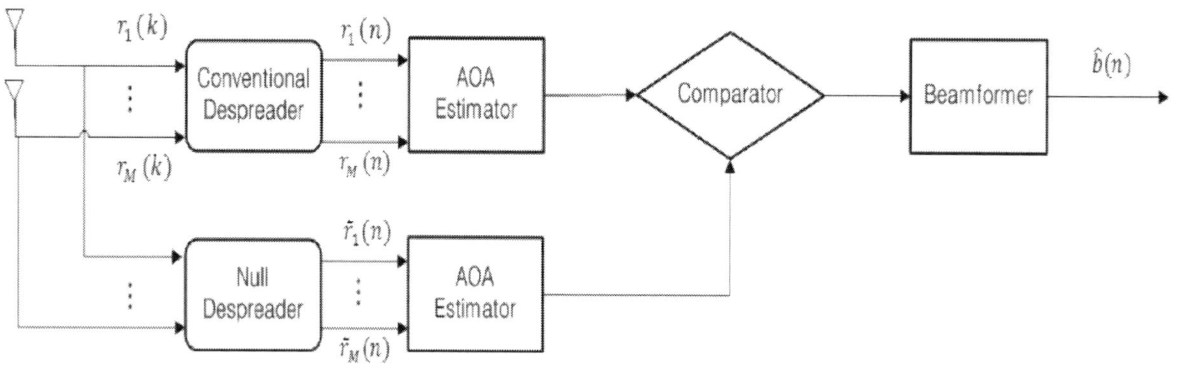

Figure 1. Architecture for the proposed GPS AOA estimation based on the null despreader

column of $\mathbf{D}$ is an AOA array response vector for the interference signal, and L is the number of interference signals. Also, we assume that the receiver utilizes a grid antenna array of size $P \times Q$ ($M = PQ$) as described in [12].

III. GPS AOA ESTIMATION BASED ON NULL DESPREADER

In this section, we describe the GPS AOA estimation technique based on the null despreader, which efficiently estimates and selects the GPS AOA when high-power interference signals are present. Figure 1 shows the system architecture for AOA estimation of the GPS signal, consisting of a conventional despreader, a null despreader, AOA estimators, and a comparator.

A. AOA Estimation Based on Conventional Despreader Output

Since each satellite employs a unique PRN code with twenty identical C/A codes, the PRN code for the i th satellite is defined as

$$\mathbf{c}_i \triangleq [\mathbf{ca}_i, \ldots, \mathbf{ca}_i]^T \tag{2}$$

where $\mathbf{ca}_i$ is a row vector (size 1023) of the C/A code for the i th satellite. The output of a conventional despreader includes the GPS signal, the interference signals, and additive noise, because the power of the GPS signal ($\mathbf{c}_i^T \mathbf{c}_i = N$) is increased above the noise power level. Since $\mathbf{c}_i^T \mathbf{c}_i = N$, it is written as

$$\begin{aligned} \mathbf{r}(n) &\triangleq \mathbf{R}(n)\mathbf{c}_i \\ &= N\mathbf{d}_c b_i(n) + \mathbf{D}\mathbf{s}(n) + \mathbf{v}(n) \end{aligned} \tag{3}$$

where $\mathbf{R}(n) \triangleq [\mathbf{r}(k), \ldots, \mathbf{r}(k+N-1)]$, $b_i(n)$ is the GPS data bit for the i th satellite, $\mathbf{s}(n) \triangleq \mathbf{S}(n)\mathbf{c}_i$,

TABLE I. SUMMARY OF RECEIVED SIGNAL MODEL IN (1)

Symbol	Size	Definition
$\mathbf{r}(k)$	$M \times 1$	Received signal vector
$\mathbf{d}_i$	$M \times 1$	Array response vector for the i th satellite
$\mathbf{D}$	$M \times L$	Array response matrix for interference signals
$\mathbf{s}(k)$	$L \times 1$	Vector of interference signals

$\mathbf{S}(n) \triangleq [\mathbf{s}(k), \ldots, \mathbf{s}(k+N-1)]$, $\mathbf{v}(n) \triangleq \mathbf{V}(n)\mathbf{c}_i$, and $\mathbf{V}(n) \triangleq [\mathbf{v}(k), \ldots, \mathbf{v}(k+N-1)]$.

For estimating AOAs of GPS and interference signals, in this paper, we consider the MUSIC algorithm which has the excellent performance for high-power signals. An auto-correlation matrix for the output signal of the conventional despreader is defined as

$$\mathbf{R}_r \triangleq E\left[\mathbf{r}(n)\mathbf{r}^H(n)\right] \tag{4}$$

and its eigenstucture is given by

$$\mathbf{R}_r \boldsymbol{\Gamma} = \boldsymbol{\Gamma}\boldsymbol{\Lambda} \tag{5}$$

where $\boldsymbol{\Lambda} = \text{diag}\{\lambda_1, \ldots, \lambda_M\}$ is a diagonal matrix, λ_l is the l th eigenvalue. Using (5), an inverse of MUSIC cost function for estimating AOAs of signals is defined as

$$\Im(\theta, \phi) \triangleq \frac{1}{\mathbf{d}^H(\theta, \phi)\boldsymbol{\Gamma}_{M-L-1}\boldsymbol{\Gamma}_{M-L-1}^H \mathbf{d}(\theta, \phi)} \tag{6}$$

where $\mathbf{d}(\theta, \phi)$ is the array response vector corresponding an elevation angle (θ) and an azimuth angle (ϕ), and $\boldsymbol{\Gamma}_{M-L-1}$ is a matrix (size $M \times (M-L-1)$) whose columns are $M-L-1$ eigenvectors corresponding to the $M-L-1$ smallest eigenvalues of $\mathbf{R}_r$. Signal AOAs are determined using the $L+1$ largest peaks of $\Im(\theta,\phi)$ in (6). The estimated results based on the conventional despreading output include AOAs of GPS and interference signals.

B. AOA Estimation Based on Null Despreader Output

A null despreader is designed to eliminate the GPS signal and its code consists of ten identical C/A codes and ten identical negative C/A codes. An example of the null despreadering code can be given by

$$\tilde{\mathbf{c}}_i \triangleq \left[\mathbf{ca}_i, -\mathbf{ca}_i, \ldots, \mathbf{ca}_i, -\mathbf{ca}_i\right]^T. \quad (7)$$

Since $\mathbf{c}_i^T \tilde{\mathbf{c}}_i = 0$, the output of the null despreader is written as

$$\begin{aligned}\tilde{\mathbf{r}}(n) &\triangleq \mathbf{R}(n)\tilde{\mathbf{c}}_i \\ &= \mathbf{D}\tilde{\mathbf{s}}(n) + \tilde{\mathbf{v}}(n)\end{aligned} \quad (8)$$

where $\tilde{\mathbf{s}}(n) \triangleq \mathbf{S}(n)\tilde{\mathbf{c}}_i$ and $\tilde{\mathbf{v}}(n) \triangleq \mathbf{V}(n)\tilde{\mathbf{c}}_i$. Defining an auto-correlation matrix for the output signal of the null despreader as

$$\mathbf{R}_{\tilde{r}} \triangleq E\left[\tilde{\mathbf{r}}(n)\tilde{\mathbf{r}}^H(n)\right] \quad (9)$$

and using processes of (5) and (6), we estimate AOAs of interference signals from the L largest peaks of $\Im(\theta,\phi)$. Since the GPS signal is removed by the null despreader, $\mathbf{\Gamma}_{M-L-1}$ is changed to $\mathbf{\Gamma}_{M-L}$ in (6). The estimated results based on the null despreading output include AOAs of only interference signals, but it does not include the GPS AOA.

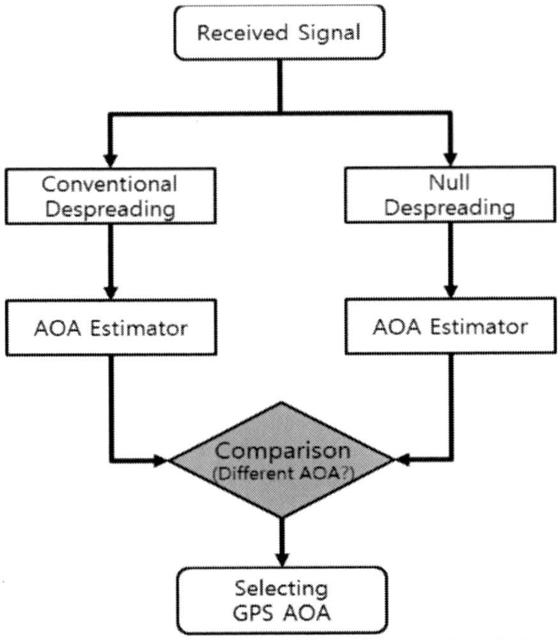

Figure 2. Flow-chart for the proposed GPA AOA estimation technique.

TABLE II. COMPUTER SIMULATION SCENARIO

Signal	Azimuth (°)	Elevation (°)	Center Frequency
GPS	27	78	-
CW	-59,2	78,78	0.16, 0.42
FM	-81	78	0.24
WB	-29,71	78,78	0.08, 0.35
Pulsed	47	78	-

C. Selection of GPS AOA

Although the estimated AOAs based on the conventional despreader consist of AOAs for interference and GPS signals, the estimated AOAs based on the null despreader consist of AOAs for only interference signals. Since the estimated AOAs of interference signals based on both despreaders are identical, we determine the AOA included in the results based on the conventional despreader but not included in the results based on the null despreader, as the GPS AOA. Figure 2 shows a flow-chart of the proposed GPS AOA estimation technique. In order to suppress interference signals, the estimated AOA information of the GPS signal is applied to the adaptive beamformer, such as MVDR or GSC.

IV. COMPUTER SIMULATION

In this section, we provide a computer simulation example to demonstrate the performance of the proposed GPS AOA estimation technique based on the null despeader, when the high-power interference signals exist. Assuming that we employ $M=8$ antenna elements at the receiver, the received signal consists of two CW interference signals, one FM interference signal, two WB noise interference signals, one pulsed interference signal with the period of 100 samples, and AWGN. The parameters of these signals are summarized in Table II. In addition, SNR of the GPS signal is -30 dB, JSR of each interference signal is 60 dB, and the modulation index and the normalized modulation frequency of the FM interferer are $\beta = 0.05$ and $f_m = 0.001$, respectively.

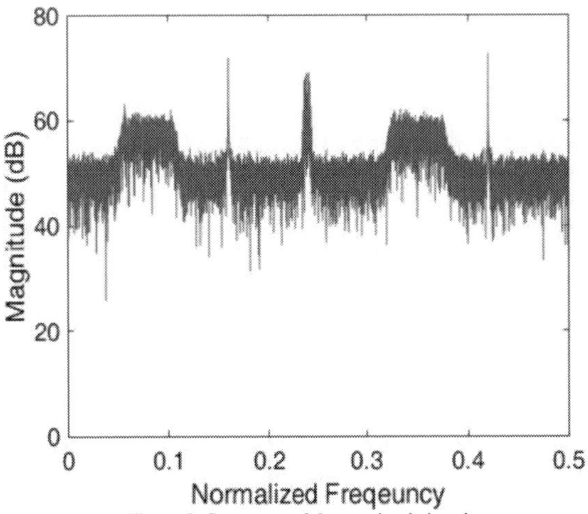

Figure 3. Spectrum of the received signal.

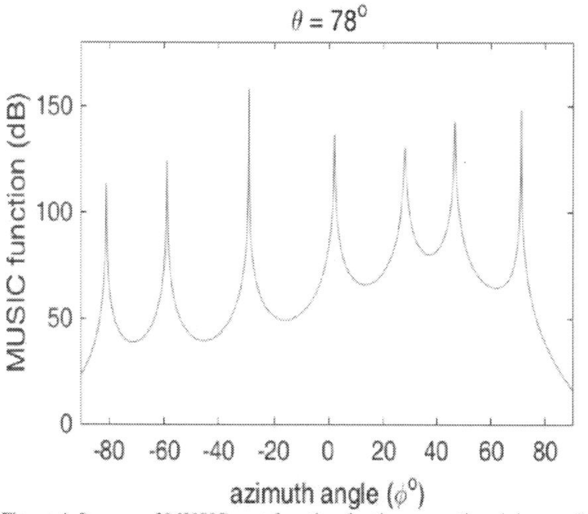

Figure 4. Inverse of MUSIC cost function for the conventional despreading output.

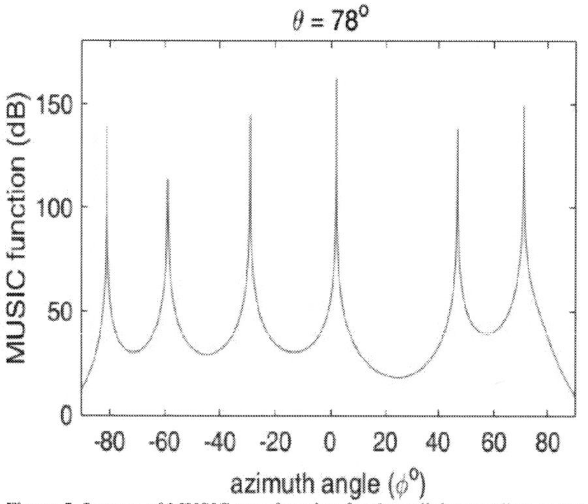

Figure 5. Inverse of MUSIC cost function for the null despreading output.

The spectrum of the received signal, which includes two CW, one FM, two WB noise, and one pulsed interferes, is shown in Figure 3. Figure 4 shows the inverse of MUSIC cost function for the conventional despreading output, at the 78 ° elevation angle, which consists of seven peaks corresponding to AOAs of one GPS signal (azimuth angle (°): 27) and six interference signals (azimuth angles (°): -81, -59, -29, 2, 48, 71). Figure 5 shows the inverse of MUSIC cost function for the null despreading output, at the 78 ° elevation angle, which consists of six peaks corresponding to AOAs of six interference signals (azimuth angles (°): -81, -59, -29, 2, 48, 71). Comparing peaks in Figure 4 and Figure 5, we select a peak, which is included in peaks of Figure 4 but excluded in peaks of Figure 5, as the GPS AOA.

V. Conclusion

The GPS AOA information is the most important factor for enhancing the quality of the GPS signal and suppressing the high-power interference signals. In this paper, we proposed the efficient GPS AOA estimation technique based on a null despreader, which is designed to remove the GPS signal but retain interference signals. Using the conventional AOA estimator such as MUSIC, the estimated result based on the conventional despreading output includes AOAs of the GPS and interference signals. However, the estimated AOA result based on the null despreading output includes AOAs of only interference signals but excludes the GPS AOA. In the proposed technique, we compare both results and select the AOA which is included in the result of the conventional despreader but is not included in the result of the null despreader, as the GPS signal. The estimated GPS AOA is applied to the adaptive beamformer for rejecting interference signals. The performance of the proposed GPS AOA estimation scheme was illustrated by the computer simulation example. The final version of the paper will include additional computer simulations and more details about algorithm.

Acknowledgment

This research was supported by Basic Science Research Program through the National Research Foundation of Korea (NRF) funded by the Ministry of Education, Science and Technology (No. NRF-2015R1D1A1A01058827).

References

[1] S. J. Vaughan-Nichols, "Will mobile computing's future be location, location, location?," Computer, vol. 42, pp. 14-17, Feb. 2009.

[2] K. Singh and M. Ismail, "OTDOA location determining technology for Universal Intelligent Positioning System (UIPS) implementation in Malaysia," Proc. 13th IEEE Int. Conf. on Networks Jointly held with the IEEE 7th Malaysia Int. Conf. on Communication, Kuala Lumpur, Malaysia, Nov. 2005, pp. 1057-1061.

[3] A. Asosheh and H. Khanifar, "A technology selection method: Hospital location detection system," 2010 5th International Symp. on Telecommunications, Tehran, Iran, Dec. 2010, pp. 992-999.

[4] E. Kaplan, Understanding GPS: Principles and Applications, Artech House, MA, USA, 1996.

[5] S. Hwang and J. J. Shynk, "Multicomponent receiver architectures for GPS inteference suppression," IEEE Trans. on Aerospace and Electronic Systems, vol. 42, no. 2, pp. 489-502, Apr. 2006.

[6] A. Ndili and P. Enge, "GPS receiver autonomous interference detection," Proc. IEEE Position Location and Navigation Symp., Palm Springs, CA, USA, Apr. 1998, pp. 123-130.

[7] J. Wang and M. G. Amin, "GPS interference cancellation performance in single and multiple MVDR beamforming methods," Proc. Fortieth Asilomar Conf. on Signals, Systems, and Computers, Pacific Grove, CA, pp. 2300-2304, Oct. 2006.

[8] D. Li, Q. Yin, P. Mu, and W. Guo, "Robust MVDR beamforming using the DOA matrix decomposition," Proc. 1st International Symposium on Access Spaces (ISAS '11), Yokohama, Japan, June 2011, pp. 105- 110.

[9] S. Hwang and J. J. Shynk, "Blind GPS receiver with a modified despreader for interference suppression," IEEE

Trans. on Aerospace and Electronic Systems, vol. 42, no. 2, pp. 503-513, Apr. 2006.

[10] W. Shao and W.-C. Wang, "A new GSC based MVDR beamformer with CSLMS algorithm for adaptive weights optimization," Proc. 4th International Congress on Image and Signal Processing (CISP '11), Shanghai, China, Oct. 2011, pp. 2299-2302.

[11] S. Hwang and J. J. Shynk, "A null despreader for interference suppression in GPS," Int. J. of Satellite Communications and Networking, vol. 29, issue 4, pp. 315-332, 2011.

[12] H. L. V. Trees, Optimum Array Processing: Part IV of Detection, Estimation, and Modulation Theory, ed., John Wiley & Sons, New York, 2002.

Design of S band Cylindrical Waveguide Slot Omnidirectional Antenna

Zhengxin Fang
East China Research Institute of Electronic Engineering
Hefei, China
e-mail:fzx1225@163.com

Abstract—**This paper presents a cylindrical waveguide slot omnidirectional antenna. The antenna has a certain number of vertical slots cut in the surface of the cylindrical waveguide to form a slot array. The vertical slots are cut at equal length. The antenna has omnidirectional radiation characteristics in the E-plane, and "8" radiation characteristics approximately in H-plane. The test results show that the antenna has broadband characteristics and the amplitude changes in the E-plane under ripple ±1 dB in S-band. The antenna has been successfully applied to products and can be used as a communication and radar antenna, among other uses.**

Keywords- cylindrical waveguide slot array; omnidirectional; broadband; antenna.

I. INTRODUCTION

The forms of the slot antenna are varied [1]. Due to the characteristics of the waveguide field distribution, the position of a single slot antenna is more flexible than the dipole antenna. Using a simple excitation method, then the slot can generate radiation [2].

Usually, in order to enhance the direction of the slot antenna, a number of slots of equal size are cut in the cylindrical waveguide sidewall by following a certain rule [3]. This becomes a cylindrical waveguide slot array antenna.

The cylindrical waveguide slot array antenna has a good mechanical strength, compact structure, high radiation efficiency, corrosion resistance, is easy to produce and has a series of other advantages, as indicated in [4]. It is widely used in the radar, microwave communications and television broadcasting systems. The cylindrical waveguide slot antenna can be widely implemented in the L, S, C, X, and other bands of horizontal polarized antenna.

In this paper, the cylindrical waveguide slot antenna is made into S band omnidirectional antenna. The test results show the antenna has an omnidirectional radiation characteristic, its amplitude changes under ripple ±1 dB in the E-plane, and has "8" radiation characteristic approximately in H-plane. It also had broadband characteristics. The design has been successfully applied to the radar and communication antennas.

The rest of the paper is structured as follows. In Section 2, we describe how to design and simulate the antenna. In Section 3, we present the test results. We conclude the paper in Section 4.

II. DESIGN AND SIMULATION

The slot array antenna can be divided into resonant array (standing wave array) and non-resonant array (traveling wave array) [2]. In the case of the standing wave of the cylindrical waveguide slot array antenna, six to seven longitudinal slots are cut around the cylindrical waveguide. The diameter φ value of the cylindrical waveguide is chosen so that the distance between the adjacent slots is about λg/2 (λg is the waveguide wavelength).The length of each slot is about λ/2 (λ is the vacuum wavelength), and the width of each slot is 2~3mm.The terminal of cylindrical waveguide is placed with metallic board for short circuit; the board is about λg/2 from the center of the terminal slots [3]. An excitation probe is provided on one side of the center of each slot. The slots of the antenna array are then excited in equal phase [4]. The harmonic waves are stimulated in cylindrical waveguide, then radiate to the outside [5]. The maximum radiation direction is perpendicular to the cylindrical waveguide surface [6]. In this paper, each slot array had six longitudinal slots in a circumference of the cylindrical waveguide. In order to obtain a higher gain, two rows of slot arrays were cut in the cylindrical waveguide. The cylindrical waveguide slot antenna structure is shown in Figure 1 and the top view of the cylindrical waveguide slot antenna is shown in Figure 2.

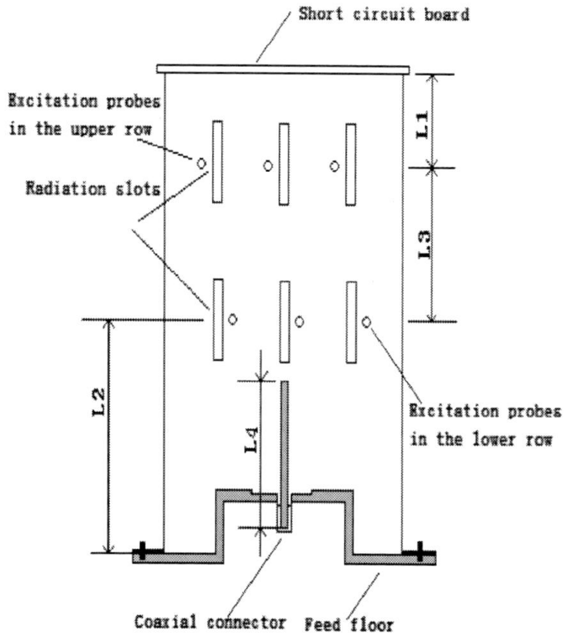

Figure 1. The cylindrical waveguide slots antenna structure diagram.

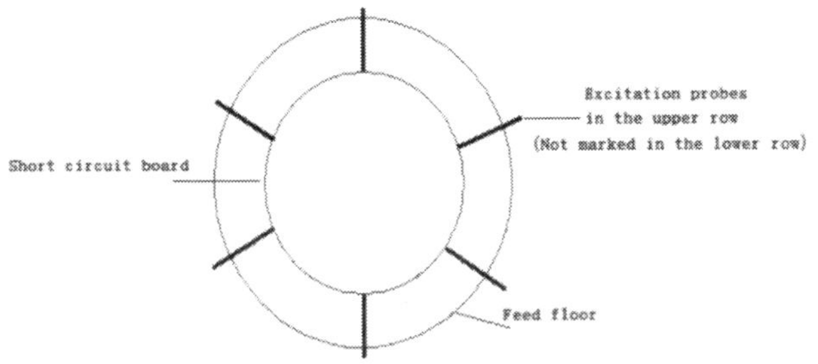

Figure 2. The cylindrical waveguide slot antenna from top view.

After the basic parameters of the antenna were calculated, the antenna was simulated and optimized by radio frequency simulation software, then given the cylindrical waveguide diameter φ= 100mm, L1 = 47mm, L2 = 143mm, L3 = 57mm, L4 = 51mm. The diameters of the probes were not obvious to the electrical performance of the antenna. The diameter of the probes was chosen to be 3mm in this paper. The experiments show that the length of the probes into the cylindrical waveguide were sensitive to the performance of the antenna. The lengths may be determined by debugging [7].

III. TEST RESULTS

The array antenna was manufactured using the above optimization calculation results. The test results showed that the lengths of the probes extending into the cylindrical waveguide had obvious influence on the performance of the antenna. The performance of the antenna was mainly changed in the standing wave and the azimuth amplitude. The changes are shown in Table I. After debugging, when the probes extended into the cylindrical waveguide with the length of 23 mm, the better electrical performance can be given. The curve of Voltage Standing Wave Ratio (VSWR) is shown in Figure 3. The far field radiation patterns of antenna in the high, medium and low frequency in E-plane, H-plane are shown in Figure 4. The antenna has omni-directional characteristic in the 360° range, its amplitude fluctuation of far field radiation pattern in E-plane does not exceed 1.9dB, and "8" radiation characteristic approximately in H-plane.

TABLE I. EFFECT OF THE LENGTH OF THE PROBE EXTENDING OUT OF THE WAVEGUIDE ON THE PERFORMANCE OF THE ANTENNA

Extending into length (mm)	The change of VSWR(Center frequency)	amplitude fluctuation in E-plane(Center frequency /dB)
15	1.36	4.2
18	1.45	3.0
23	1.52	1.9
27	1.89	1.7
30	2.21	1.5

From Table I, when the probes extended into the cylinder waveguide with the lengths of 23 mm, the values of electrical performance were better.

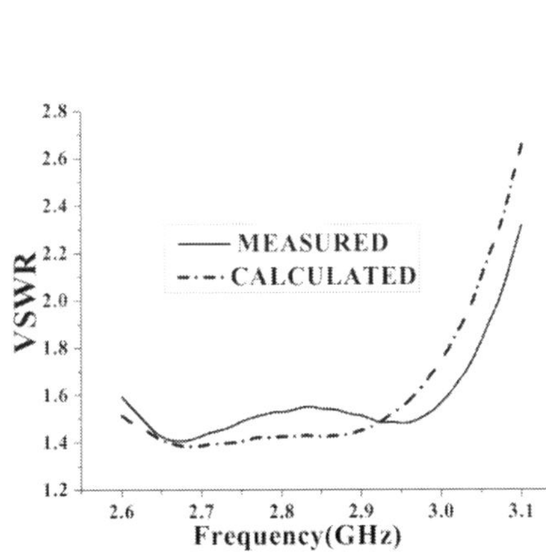

Figure 3. The curve of VSWR by measured and calculated

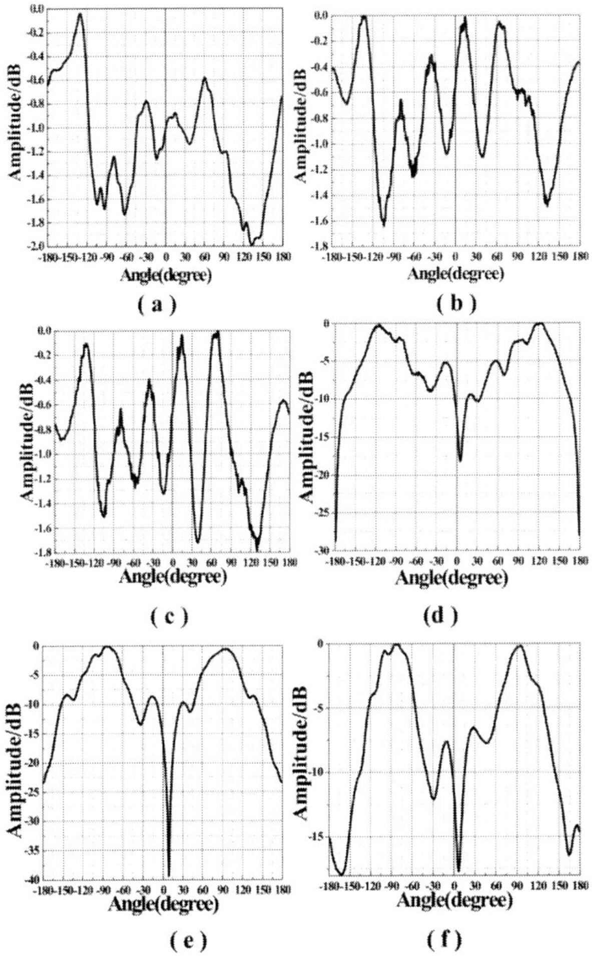

Figure 4. Measured radiation pattern.

(a) E-plane at 2.6GHz. (b) E-plane at 2.8GHz. (c) E-plane at 3.0GHz. (d) H-plane at 2.6GHz. (e) H-plane at 2.8GHz. (f) H-plane at 3.0GHz.

As seen in Figure 3, when the lengths of the probes in the cylinder waveguide were 23 mm, the values of VSWR were smaller. The value of VSWR can be less than 1.6 in 2.6 ~ 3.0 GHz frequency range.

IV. CONCLUSION

In this paper, an S band cylindrical waveguide slot omnidirectional antenna was presented. The antenna had an omnidirectional radiation characteristic in the E-plane and an "8" radiation characteristic approximately in H-plane. It had broadband characteristics and its amplitude changed under ripple ±1 dB in the E-plane. It can be used as communication and radar antenna.

REFERENCES

[1] R. S. Elliott "Antenna Theory and Design.Englewood Cliffs," NJ Prentice Hall, 1981.

[2] C. Jan and P. Hsu, "Moment Method Analysis of Sidewall Inclined Slots in Rectangular Waveguides," IEEE Trans. Antennas Propagat., vol. 39, pp. 68-73,1991.

[3] W. Wang, J. Jin, and S. Zhong, "Wide-band diaphragm incentive of waveguide narrow side not inclined slot array antenna," Journal of Microwaves, vol. 21, pp. 30-33,2005.

[4] Z. Fang and J. Lu, "Broad bandwidth pressure Angle narrow rectangular waveguide antenna study," Radar & Ecm, vol. 2, pp. 16-17,2009.

[5] C. Ma, H. Lang and Z. Chen, "Broadband low side-lobe single ridged waveguide slot array antenna design," Journal of Microwaves, vol. 25,pp. 224-227, 2010.

[6] G. A. Casula, G. Mazzarella and G. Montisci, "Design of shaped beam planar arrays of waveguide longitudinal slots," International Journal of Antennas and Propagation, 2013, Article ID 767342.

[7] H. Chen, B. F. He and Y. X Zhang,"Study on large angle slotted waveguide antenna with low sidelobe," Journal of Microwaves, vol.30 (S1), pp. 283-286, 2014.

Design of an UWB Meter-wave Oblique Polarized Array Antenna

Jia Fang, Hao Qin, Yongdong Zang
ECRIEE
East China Research Institute of Electronic Engineering
Hefei, China
fjftt@mail.ustc.edu.cn,qinhao@163.com,zyd123@163.com

Abstract—**According to the airborne meter-wave electronic reconnaissance and interference need, an 150-350MHz 45deg oblique polarized array antenna is designed. The size of the log-periodic antenna is reduced by make fractal to the ordinary log-periodic antenna dipoles and load to the top, which is like a tree. The height of the antenna is also reduced, and the feed parallel lines are wrapped around by dielectric materials. The array is simulated and optimized, and the results show the array can achieve airborne meter-wave broadband wide-angle scan work.**

Keywords- ultral-wideband; log-periodic antenna; fractal; 45deg oblique polarized.

I. INTRODUCTION

In recent years, because of the advantages of finding anti-stealth targets and counterwork anti-radiation missiles, the meter-wave radar has been developed greatly. So, the need of meter-wave radar counter equipment is also growing [1].

Being a popular ultra-broadband antenna, the log-periodic antenna has a lot of advantages like technology maturation, good acclimatization, high power capacitance [2][3], which are suitable for the electronic warfare equipment use. However, the ordinary log-periodic antenna has a large size, which is bigger than the element spacing of a broadband wide-angle scan array, and also, it is not suitable for the airborne platform use. There is a need to miniaturize the design.

According to the airborne meter-wave broadband wide-angle scan antenna array need, the transverse size and longitudinal size of the antenna are reduced by make fractal to the ordinary log-periodic antenna dipoles, load the top and wrap around the feed parallel lines with dielectric materials. A 150-350MHz 45deg oblique polarized antenna array is designed.

II. ANTENNA ELEMENT

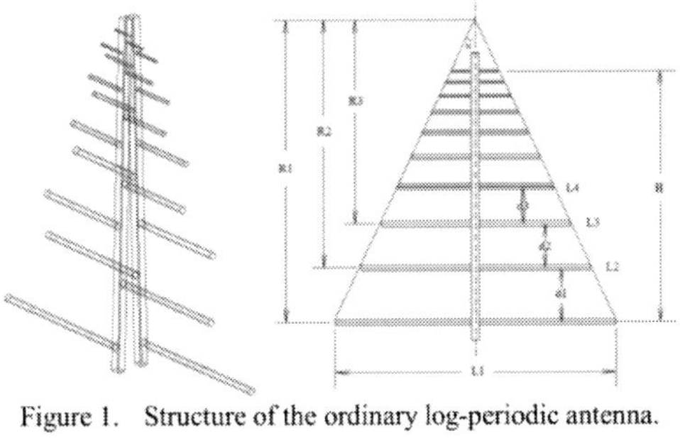

Figure 1. Structure of the ordinary log-periodic antenna.

The ordinary log-periodic antenna is composed of a pair feed parallel lines and some pair radiation dipoles, as shown in Figure 1 [4]. According to the design formula, the size of the ordinary log-periodic antenna is determined by the ratio factor τ, the interval factor σ, and the minimum work frequency [5]. For 150-350MHz work, the high of the parallel lines is not less than 0.8m, and the length of the longest dipole is about 1m. But for the 350MHz scan 45deg work, the element spacing of the antenna array must be less than 0.45m. When the ordinary log-periodic antenna is used for the array, there is too much overlap of the dipoles, which will seriously deteriorate the cross-polarization, the coupling, the sideline level, the gain and other indicators of the antenna. The 0.8m height of the antenna can not meet the installation requirements of the airborne platform.

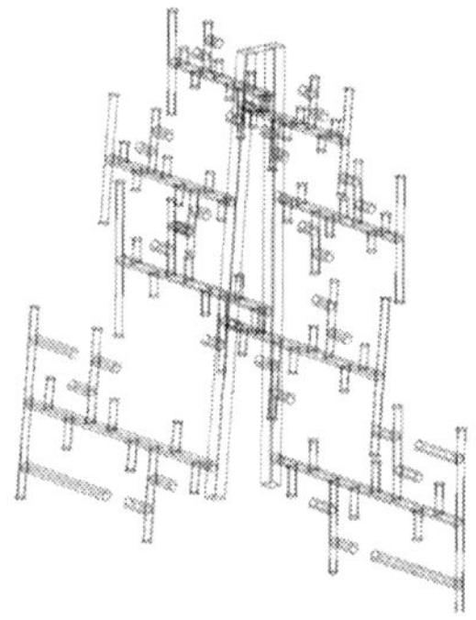

Figure 2. Structure of the tree antenna.

The tree antenna is composed of feed parallel lines, fractal dipoles, and metal sticks, which are loaded to the top of dipoles, as shown in Figure 2. Two metal sticks are added to the dipole in the two thirds of the normal dipole, and the length of each one is 1/3 of the normal dipole, as 19-22 metal sticks shown in Figure 3, which is called the first order fractal dipole. In the same way, we do the fractal again to each section of the first order fractal dipole, which is called the second order fractal dipole, as shown in Figure 3. The length of the dipoles are reduced by the fractal structure, and the lowest work frequency is further reduced by load metal sticks to the top of dipoles, which make the transverse size of the antenna drop to half of ordinary log-periodic antenna. So, the antenna could be the element of an antenna array with scan ±45deg. The length of the feed parallel lines, which is also the height of the antenna, is also reduced to be 53% of the ordinary log-periodic antenna, which is suitable for the airborne platform use.

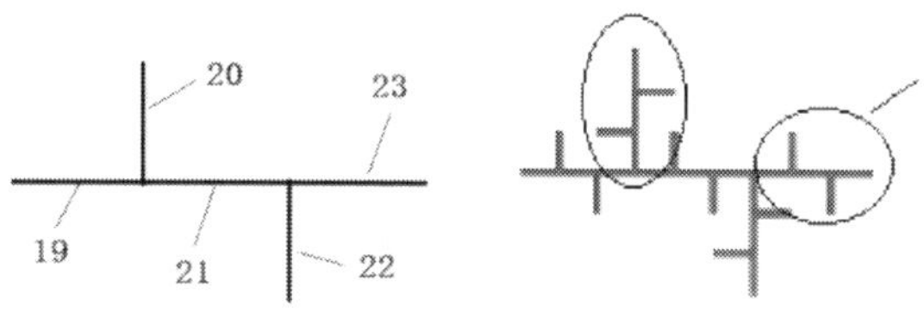

Figure 3. Structure of the first order and second order fractal dipole.

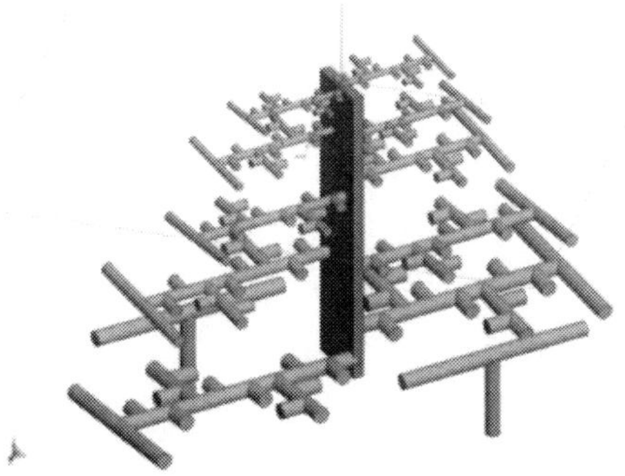

Figure 4. Simulation model of the tree antenna.

The antenna element is simulated and optimized by FEKO software, as shown in Figure 4. The feed parallel lines are wrapped around by dielectric materials, to reduce the antenna port impedance to be easy match, which serve also to support weight. After being optimized, the height of the antenna is 400mm, and the size is 700mm×300mm. Figure 5 shows the Voltage Standing Wace Ratio (VSWR) results of the antenna, and its blow 2 at whole work band. Figure 6 is the normal gain of the antenna.

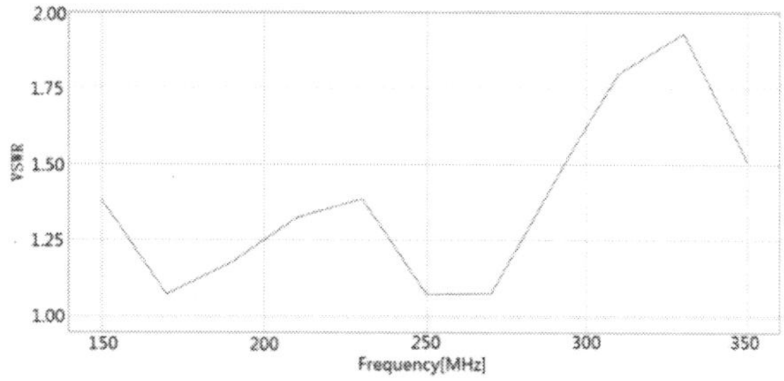

Figure 5. VSWR of the antenna element.

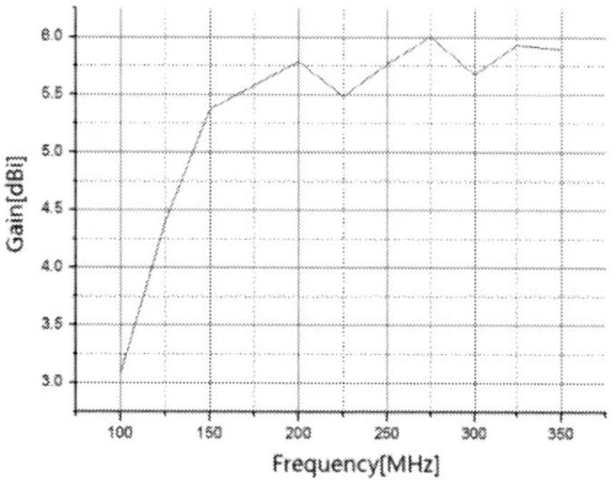

Figure 6. Normal gain of the antenna element.

III. ARRAY ANTENNA

The antenna array is simulated by FEKO software, as shown in Figure 7. The size of the array is 2×6. For 350MHz scan 45deg work with no grating lobe, the horizontal element spacing chooses 400mm. There is no scan in vertical direction, so the vertical spacing could be bigger and is 700mm.

In order to improve the low frequency performance of the antenna array, some parasitic dipoles are added between each antenna elements, and the height is lower than the lowest dipole of the antenna, as shown in Figure 7. The length and height of the parasitic dipole are 270mm and 120mm.

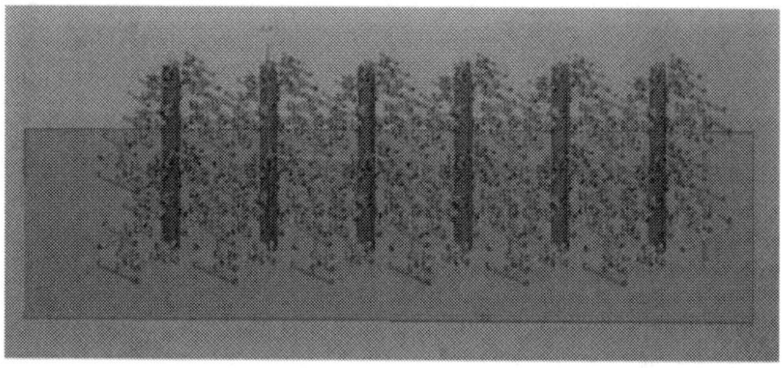

Figure 7. Simulation model of the array antenna.

The antennas are re-optimized. Figure 8 shows the VSWR results of the array, and all are blow 2.4 at whole work band. Figure 9 is the normal gain of the array, and it is bigger than 11.5dBi. Figure 10 shows the lobe pattern of the array at horizontal direction at 150MHz, 250MHz and 350MHz without scan, and Figure 11 shows scan 45deg.

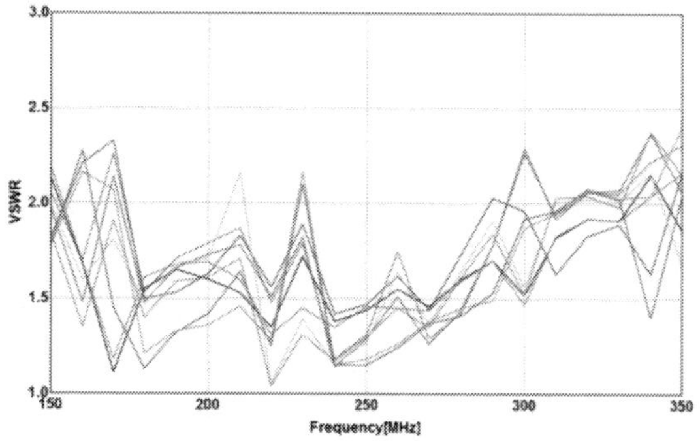

Figure 8. VSWR of the array.

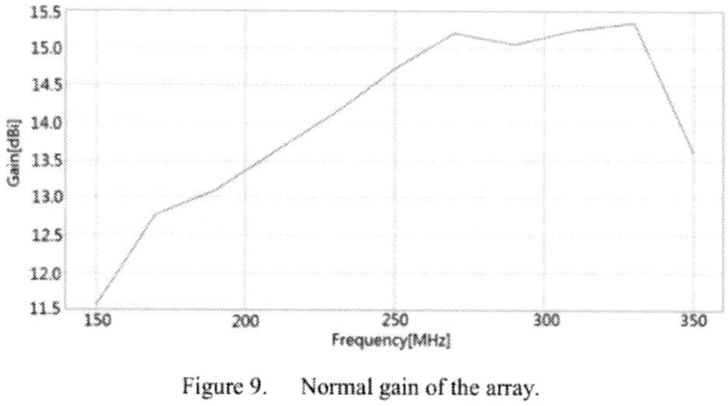

Figure 9. Normal gain of the array.

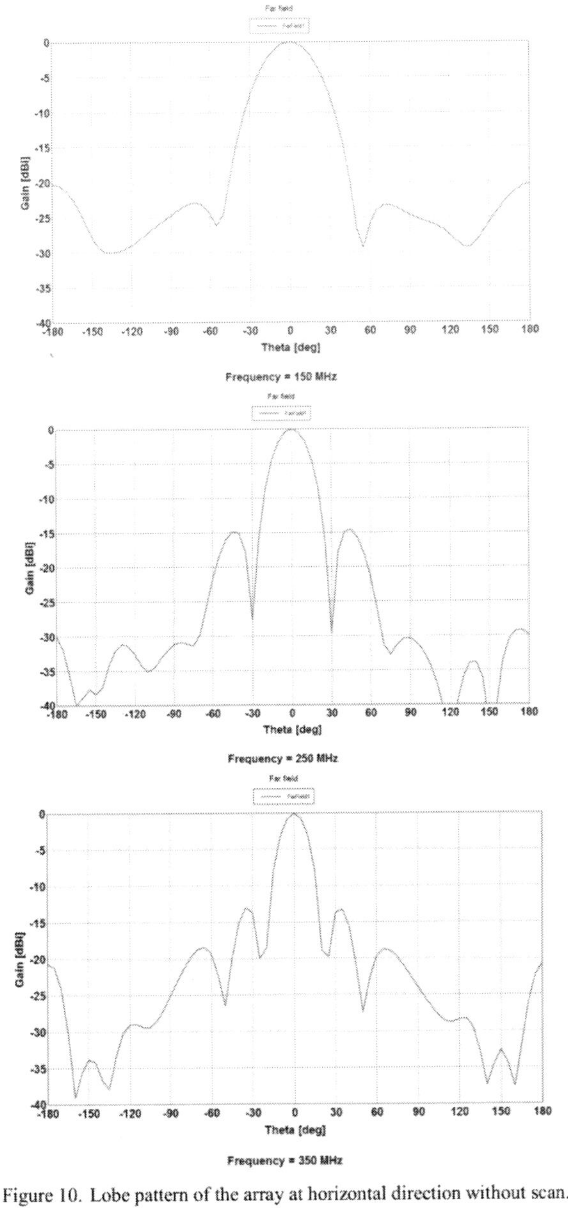

Figure 10. Lobe pattern of the array at horizontal direction without scan.

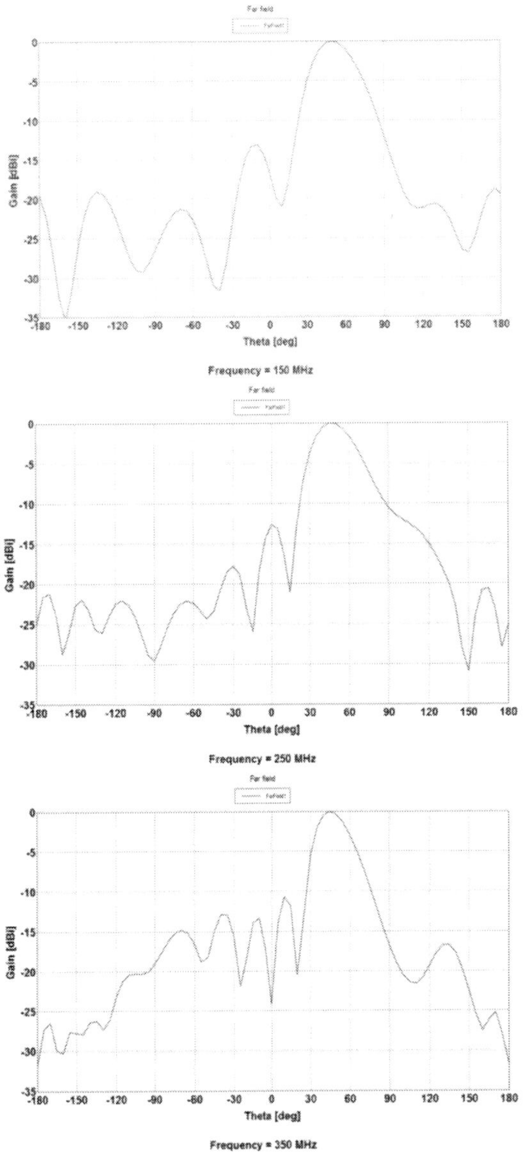

Figure 11. Lobe pattern of the array at horizontal direction with scan 45deg

IV. CONCLUSION

By make fractal to the ordinary log-periodic antenna dipoles, load to the top and wrap around the feed parallel lines with dielectric materials, the transverse size and height of the antenna are half 53% of the ordinary log-periodic antenna, which could achieve the needed size for the airborne meter-wave broadband wide-angle scan antenna. The antenna array is simulated and optimized by FEKO software, and the results showed it meet the needs of the system work.

REFERENCES

[1] R. A. Poisel, "Antenna Systems and Electronic Warfare Applications", published by Artech House, INC, 2012

[2] D. E. Anagnostou, "A Small Planar Log- periodic Koch- pipole Antenna," IEEE Antennas and Propagation Society International Symposium 2006, Albuquerque, 2006.

[3] A. A. Gheethan, "Log- periodic Dipole Arrays," International Symposium 2008, San Diego, CA, USA, July 5- 12,2008.

[4] X. J. Kang, "Antenna Theory and Design", Beijing, National Defense Industry Press, 1995.

[5] C. L. Yuan, "Ultra-broadband Antenna Theory and Technology", Haerbin, Haerbin Industry University Press, 2006.

Outage Probability of Triple-Hop Mixed RF/FSO/RF Stratospheric Communication Systems

Emmanouel T. Michailidis[1,2], Nikolaos Nomikos[3], Petros Bithas[1,2], Demosthenes Vouyioukas[3], and Athanasios G. Kanatas[1]

[1]Department of Digital Systems, University of Piraeus, Piraeus, Greece
e-mail: {emichail, pbithas, kanatas}@unipi.gr

[2]Department of Electrical and Electronics Engineering, University of West Attica, Campus 2, Aigaleo, Athens, Greece

[3]Department of Information and Communication Systems Engineering, University of the Aegean, Karlovassi, Samos, Greece
email: {nnomikos, dvouyiou}@aegean.gr

Abstract—**This paper proposes a triple-hop mixed Radio-Frequency/Free-Space-Optical/Radio-Frequency (RF/FSO/RF) communication system, which intends to support wireless long-range links between two terrestrial stations via two stratospheric relays. It is considered that these terrestrial stations communicate with the relays over RF links, whereas the relays communicate with each other over a FSO link. The RF channels experience Rician fading due to the Line-of-Sight (LoS) and Non-Line-of-Sight (NLoS) signal components. Besides, the optical channel is affected by atmospheric attenuation, atmospheric turbulence, and pointing errors. Mathematical expressions for the outage probability are derived, considering the beam wander effect. The results demonstrate the theoretical derivations.**

Keywords-Atmospheric turbulence; beam wander; Free-Space-Optical (FSO) communications; High-Altitude Platforms (HAPs); outage probability; pointing errors; Rician channels.

I. Introduction

The development of next-generation wireless broadband communications comprises the seamless integration of terrestrial and aerospace infrastructures over heterogeneous networks [1]. In recent years, the use of mobile airborne services via High-Altitude Platforms (HAPs) flying in the stratosphere has been suggested for civil and military applications [2]. To successfully fulfill the demands of future wireless communication services, shifting from the Radio-Frequency (RF) domain to the optical domain is indispensable [3] [4]. In Free-Space-Optical (FSO) systems, the data is transmitted at high data rates in the multigigabit regime using collimated highly directed laser beams with compact equipment and low power consumption. The FSO technology ensures privacy with low probability of interception, immunity to electromagnetic interference, and exemption from spectrum regulatory restrictions.

The stratosphere stands for a propagation medium well suited for FSO connections. Hopefully, the inter-HAP crosslink is above attenuating clouds, fog or significant aerosols. Hence, inter-HAP distances of up to 900 km are possible with 100% availability [5]. Nevertheless, stratospheric FSO links may experience substantial fluctuations in both the intensity and the phase of the received signal due to variations of the Index of Refraction Turbulence (IRT) along the propagation path resulting by the inhomogeneities in temperature and pressure fluctuations, especially for link distances of 1 km and above [6]. Moreover, beam pointing error effects are highly pronounced on the performance of a stratospheric network, which involves large distances and possible displacement of the platforms due to the winds or pressure variations of the stratosphere [7]. In addition, both mechanical vibrations and electronic noise may cause the received spots from the laser beams to wander on the detector planes. Hence, acquisition, pointing, and tracking errors should also be mitigated to achieve high directivity of the transmitted beam. To extend the network range, and effectively combat the effect of turbulence, the use of relay-aided transmission schemes has been suggested [8]. In [7], a full-FSO system with multiple stratospheric relays and fixed stations on the ground was proposed. However, FSO systems should operate under accurate pointing requirements. Hence, employing FSO techniques is highly challenging and insufficient as soon as the terrestrial stations are in motion.

Motivated by these observations, this paper investigates the use of two stratospheric relays acting as Decode-and-Forward (DF) relay nodes, in order to facilitate challenging triple-hop long-range wireless communication between two terrestrial mobile stations. Although the Amplify-and-Forward (AF) scheme is simple to implement, DF offers better performance and is a common assumption in multi-hop networks. A mixed RF/FSO/RF airborne communication scenario is considered, where an RF transmission scheme is adopted for the communication between the terrestrial stations and the stratospheric relays, whereas these relays are interconnected through an FSO link. A stratospheric communication channel is expected to be Rician in its general form [9]. Hence, the Rician distribution is adopted to model the channels for the RF links. To investigate the effect of atmospheric turbulence on the FSO link, a Kolmogorov power spectrum for refractive-index fluctuations is considered and the Gamma-Gamma (G-G) distribution is used, which matched data values obtained from measurements, under a variety of turbulence conditions [10]. Finally, the effects of pointing errors are described by the Rayleigh distribution [7]. Mathematical expressions are derived for the outage probability and numerical results are provided for different values of the system parameters.

The rest of the paper is organized as follows. Section II presents the system model. In Section III, the RF and FSO fading channels are modeled. In Section IV, the outage probability is mathematically derived. Results are provided in Section V. Finally, conclusions are drawn in Section VI.

II. System Model

This paper considers a mixed RF/FSO/RF airborne communication system with slowly-varying and frequency-flat-fading channels. As shown in Figure 1, the transmitted signal from a source node propagates through two serial DF stratospheric relays of similar size and type before arriving at a destination node. It is considered that the direct link between the source and the destination is obstructed due to high attenuation. To aid our analysis, the subscripts S, D, and R_m, where $1 \leq m \leq 2$, are affiliated with the source, the destination, and the m-th relay, respectively. In this system, the S–R_1 and R_2–D links use RF technology, while the R_1–R_2 link is based on FSO technology. The communication is assumed to operate in a half-duplex mode and to be conducted over three phases: $S \rightarrow R_1$, $R_1 \rightarrow R_2$, and $R_2 \rightarrow D$. The source and the destination are equipped with single antennas. Besides, the relays not only include antennas, but also lasers and photo-detectors.

A. First RF Link

The received signal at the stratospheric relay R_1 can be expressed as [11]

$$y_{S,R_1} = \sqrt{P_S} h_{S,R_1} x_{S,R_1} + n_{R_1}, \tag{1}$$

where P_S is the transmit power, h_{S,R_1} is a non-zero-mean complex Gaussian random variable, x_{S,R_1} is the transmitted symbol from the source with $\mathrm{E}\left\{\left|x_{S,R_1}\right|^2\right\} = 1$, $\mathrm{E}\{\cdot\}$ is the statistical expectation operator, n_{R_1} represents the zero-mean complex Gaussian noise at R_1 with N_{o,R_1} noise Power Spectral Density (PSD). Using (1), the instantaneous Signal-to-Noise Ratio (SNR) at R_1 can be written, respectively, as

$$\gamma_{S,R_1} = \frac{P_S}{N_{o,R_1}} \left|h_{S,R_1}\right|^2. \tag{2}$$

B. FSO Inter-HAP Link

In the FSO link, the cost-effective Intensity Modulation Direct-Detection (IM/DD) is employed. The received RF signal at the R_1 is decoded and converted to optical signal by employing the Subcarrier Intensity Modulation (SIM) technique. In particular, after filtering by a Bandpass Filter (BPF), a Direct Current (DC) bias is added to the filtered RF signal to ensure that the optical signal is non-negative. Then, the biased signal is sent to a continuous wave laser driver. The retransmitted optical signal at R_1 is written as [12]

$$y_{R_1,opt} = \sqrt{P_{opt}} \left(1 + M y_{S,R_1}\right), \tag{3}$$

where P_{opt} denotes the average transmitted optical power and it is related to the relay electrical power P_{R_1} by the electrical-to-optical conversion efficiency η_1 as $P_{opt} = n_1 P_{R_1}$, and M denotes the modulation index.

The optical signal at R_2 received from R_1 can be expressed as

$$y_{R_1,R_2} = I\left[\sqrt{P_{opt}} \left(1 + M y_{S,R_1}\right)\right] + n_{R_2}, \tag{4}$$

where $I > 0$ is the received fading gain (irradiance) between the laser of R_1 and the photodetector of R_2 through the optical channel and n_{R_2} is the zero-mean complex Gaussian noise at R_2 with N_{o,R_2} noise PSD.

The DC component is filtered out at R_2 and an optical-to-electrical conversion is performed. Then, assuming that $M = 1$, the received signal can be expressed as

$$y_{R_1,R_2} = I\sqrt{P_{ele}}\left[\sqrt{P_{opt}} y_{S,R_1}\right] + n_{R_2}, \tag{5}$$

where $P_{ele} = n_2 P_{opt} = n_1 n_2 P_{R_1}$ is the electrical power received at R_2 and n_2 is the optical-to-electrical conversion efficiency.

The instantaneous SNR at R_2 can be approximated as [11]

$$\gamma_{R_2} \approx \min\left\{\gamma_{S,R_1}, \gamma_{R_1,R_2}\right\}, \tag{6}$$

where $\gamma_{R_1,R_2} = \left(n_1 n_2 P_{R_1} I^2\right) / N_{o,R_2}$.

The performance of the FSO link is limited by background radiation and thermal noise, which can be modeled as independent and identically distributed (i.i.d.) AWGN, as an accurate approximation of the Poisson photon-counting detection model [13].

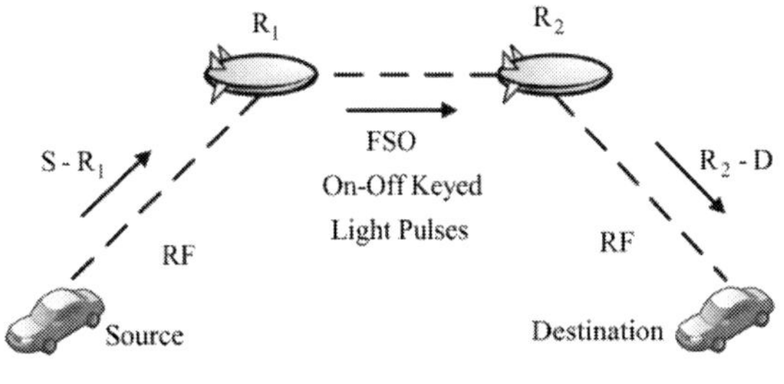

Figure 1. Simple representation of the triple-hop mixed RF/FSO/RF stratospheric communication system.

C. Second RF Link

The FSO signal at the R_2 is finally reconverted to an RF signal before being sent to the destination over the second RF link. The signal received at the destination is written as

$$y_D = \sqrt{P_{R_2}} h_{R_2,S} x_{R_2,S} + n_D, \quad (7)$$

where P_{R_2} is the transmit power, h_{R_2S} is a non-zero-mean complex Gaussian random variable (leading to a Rician distributed amplitude), $x_{R_2,S}$ is the signal transmitted by the R_2 with $\mathrm{E}\left\{\left|x_{R_2,S}\right|^2\right\} = 1$, and n_D is the zero-mean complex Gaussian noise at D with $N_{o,D}$ noise PSD. Using (7), the SNR at the destination can be written as

$$\gamma_{R_2,D} = \frac{P_{R_2}}{N_{o,D}} \left|h_{R_2,D}\right|^2. \quad (8)$$

The instantaneous end-to-end SNR at the destination can be approximated as [11]

$$\gamma_D \approx \min\left\{\gamma_{S,R_1}, \gamma_{R_1,R_2}, \gamma_{R_2,D}\right\}. \quad (9)$$

III. STATISTICAL MODELING OF THE TRIPLE-HOP MIXED RF/FSO/RF CHANNEL

In this paper, the Rician distribution is utilized to model the channels for the RF links. The Probability Density Function (PDF) of the instantaneous SNR received at R_1, denoted as γ_{S,R_1}, is given by [14]

$$f_{\gamma_{S,R_1}}\left(\gamma_{S,R_1}\right) = \frac{K_1+1}{\bar{\gamma}_{S,R_1}} \exp\left(-\left(K_1+1\right)\frac{\gamma_{S,R_1}}{\bar{\gamma}_{S,R_1}} - K_1\right) \times I_0\left(2\sqrt{K_1\left(K_1+1\right)\frac{\gamma_{S,R_1}}{\bar{\gamma}_{S,R_1}}}\right), \quad (10)$$

where K_1 is the Rician factor of the first RF link, i.e., the average power ratio of the Line-of-Sight (LoS) component to the Non-Line-of-Sight (NLoS) component, and $I_0(\cdot)$ is the modified Bessel function of the first kind and zero-th order. The Rician factor K_1 strongly depends on the elevation angle of the HAPs and the operating frequency [15]. The PDF $f_{\gamma_{R_2,D}}\left(\gamma_{R_2,D}\right)$ of the Rician distribution for the second RF link with Rician factor K_2 can be defined as in the first RF link by replacing the indices.

For the FSO link, a composite optical channel model is used [7]. In particular, the channel state I models the random attenuation of the propagation channel that arises due to the path loss I^l, the atmospheric turbulence I^a, and the pointing errors I^p. The combined optical channel model is defined as follows

$$I = I^l I^a I^p. \quad (11)$$

Since the time scales of these fading processes are of the order of 10^{-3}s to 10^{-2}s [16], which are far larger than the bit interval ($\approx 10^{-9}$s for multi-Gbps systems), I is constant over a large number of transmitted bits [17].

A. Effects of Atmospheric Attenuation

The atmospheric attenuation is deterministic and is defined by the exponential Beers–Lambert Law [18]

$$I^l(L) = \exp(-\sigma L), \quad (12)$$

where $I^l(L)$ is the atmospheric attenuation loss of the FSO link over a path of length L, i.e., the distance between the relays, and σ is the wavelength- and weather-dependent attenuation coefficient. The attenuation I^l is considered fixed during a long period of time.

B. Effects of Turbulence

In this paper, the effects of turbulence are modeled using the G-G distribution. Hence, the PDF of I^a is given by [19]

$$f_{I^a}\left(I^a\right) = \frac{2(ab)^{(a+b)/2}}{\Gamma(a)\Gamma(b)}\left(I^a\right)^{\frac{(a+b)}{2}-1} K_{a-b}\left(2\sqrt{abI^a}\right), \quad (13)$$

where $K_u(\cdot)$ is the u-th order modified Bessel function of the second kind. Note that (13) can be written in terms of the Meijer's G-function as $K_u(x) = 0.5G_{0,2}^{2,0}\left[x^2/4\middle|u/2\ u/2\right]$ [20]. The parameters a and b are directly related to the atmosphere and can be adjusted to achieve a good agreement between $f_{I^a}\left(I^a\right)$ and measurement data. These parameters are determined by the Rytov scintillation model. However, for large distances, as those involved in the proposed system, this model is inappropriate due to the hot spot displacement of the beam caused by the beam wander, i.e., the deflection of the beam due to turbulence. By considering the beam wander effect, a and b can be modeled as [21]

$$a = \left[\frac{4.42\sigma_R^2 \Lambda_e^{5/6}\sigma_{pe}^2}{w_{LT}^2} + \exp\left(\frac{0.49\sigma_B^2}{\left(1+0.56(1+\Theta)\sigma_B^{12/5}\right)^{7/6}}\right) - 1\right]^{-1}, \quad (14)$$

$$b = \left[\exp\left(\frac{0.51\sigma_B^2}{\left(1+0.69\sigma_B^{12/5}\right)^{5/6}}\right) - 1\right]^{-1}, \quad (15)$$

where $\sigma_R^2 = 1.23C_n^2 k^{7/6} L^{11/6}$ is the Rytov variance, C_n^2 is the index of refraction structure parameter, $\Lambda_e = 2L/kw_{LT}^2$,

$k = 2\pi / \lambda$ is the wave number, λ is the carrier wavelength, $w_{LT} = w\sqrt{1+1.33\sigma_R^2 \Lambda^{5/6}}$ is the long-term spot radius, w is the beam radius at the receiver, $\Lambda = 2L/kw^2$, σ_{pe}^2 is the variance of pointing error caused by beam wander ([22], p. 350), σ_B^2 is the Rytov variance with beam wander correction ([22], p. 351), $\Theta = \Theta_0 / \left(\Theta_0^2 + \Lambda_0^2\right)$, $\Theta_0 = 1 - L/F_0$, F_0 is the phase curvature parameter of the Gaussian beam at the transmitter, $\Lambda_0 = 2L/kw_0^2$, and w_0 is the transmitter beam radius. Using [23, eq. (3)], one can also obtain the Scintillation Index (SI) σ_I^2. The value of C_n^2 represents the turbulence condition of the atmosphere and can be estimated using the Hufnagel-Valley model, which is based on various empirical scintillation data of the atmosphere. This model defines the index of refraction structure parameter as a function of the wind speed and the altitude of the HAPs above the ground level as ([22], p. 481)

$$C_n^2(h) = 0.00594\left(\frac{u}{27}\right)^2 \left(10^{-5}h\right)^{10} \exp\left(-\frac{h}{1000}\right) + 2.7\times10^{-16}\exp\left(-\frac{h}{1500}\right) + \hat{A}\exp\left(-\frac{h}{100}\right), \quad (16)$$

where $\hat{A}$ is the nominal value of the $C_n^2(0)$ at the ground, u is wind speed in m/s, and h is the altitude of the HAPs in meters. As C_n^2 increases, the turbulence becomes stronger. For applications involving propagation along a horizontal path, it is customary to assume that C_n^2 remains constant [22].

C. *Effects of Pointing Errors*

To ensure that the communication scenario is viable, proper LoS alignment between the relays via appropriate pointing and tracking mechanisms is required. In this paper, a model that incorporates geometric spreading in the pointing error PDF ([18], p. 1703) is used. Considering a Gaussian beam profile of beam waist w_z (at a distance z) and modeling the random radial displacement of the beam at the detector as Rayleigh distributed, the PDF of the pointing error I^p includes the random attenuation due to both geometric spreading and pointing errors and can be expressed as [18]

$$f_{I^p}\left(I^p\right) = \frac{\gamma^2}{(A)^{\gamma^2}}\left(I^p\right)^{\gamma^2-1}, \quad 0 \le I^p \le A, \quad (17)$$

where $\gamma = w_{eq}/2\sigma_s$, $w_{eq}^2 = w_z^2\sqrt{\pi}\cdot \mathrm{erf}(u) / \left[2u\exp\left(-u^2\right)\right]$ is the equivalent beam width at the receiver, $u = \left(\sqrt{\pi}a_r\right)/\left(\sqrt{2}w_z\right)$, σ_s is the jitter standard deviation due to the pointing error at the detector determined by the degree of the misalignment between the apertures, $A = \left[\mathrm{erf}(u)\right]^2$, and a_r is the receiver aperture radius.

Using (12), (13), and (17), the PDF of the composite channel in (11) can be evaluated by writing the Bessel function $K_u(\cdot)$ in terms of the Meijer's G-function as [7]

$$f_I(I) = \frac{ab\gamma^2}{AI^l\Gamma(a)\Gamma(b)} G_{1,3}^{3,0}\left[\frac{ab}{AI^l} I \,\middle|\, \begin{matrix} \gamma^2 \\ \gamma^2-1, a-1, b-1 \end{matrix}\right]. \quad (18)$$

Using (18), the PDF of the instantaneous SNR received at R_2, denoted as γ_{R_1,R_2}, can be expressed as

$$f_{\gamma_{R_1,R_2}}\left(\gamma_{R_1,R_2}\right) = \frac{ab\gamma^2}{2AI^l\Gamma(a)\Gamma(b)}\sqrt{\frac{1}{\gamma_{R_1,R_2}\overline{\gamma}_{R_1,R_2}}} \times G_{1,3}^{3,0}\left[\frac{ab}{AI^l}\sqrt{\frac{\gamma_{R_1,R_2}}{\overline{\gamma}_{R_1,R_2}}} \,\middle|\, \begin{matrix} \gamma^2 \\ \gamma^2-1, a-1, b-1 \end{matrix}\right], \quad (19)$$

where $\overline{\gamma}_{R_1,R_2}$ is the average SNR received at R_2.

IV. Derivation of Outage Probability

The outage probability is defined as the probability that the SNR at the destination falls below a predetermined outage threshold γ_{out}, i.e., $P_{out} = \Pr\left[\gamma_D \le \gamma_{out}\right]$, where $\Pr[\cdot]$ is the probability operation. In this case, the communication system cannot achieve adequate reception. The outage probability can be obtained from the Cumulative Distribution Function (CDF) of the end-to-end SNR as $P_{out} = F_{\gamma D}\left(\gamma_{out}\right)$. This CDF can be written in terms of the CDFs of the three hops' SNRs as [24]

$$F_{\gamma D}\left(\gamma_{out}\right) = 1 - \left[\left(1 - F_{\gamma S,R_1}\left(\gamma_{out}\right)\right)\left(1 - F_{\gamma R_1,R_2}\left(\gamma_{out}\right)\right)\left(1 - F_{\gamma R_1,D}\left(\gamma_{out}\right)\right)\right], \quad (20)$$

where $F_{\gamma S,R_1}\left(\gamma_{out}\right)$, $F_{\gamma R_1,R_2}\left(\gamma_{out}\right)$, and $F_{\gamma R_1,D}\left(\gamma_{out}\right)$ are the CDFs of the SNRs of the first, second, and third hop, respectively. One observes that the system falls in outage providing that at least one of the three hops gets in outage or, equivalently, the SNR of one hop becomes less than γ_{out}.

The CDF of the instantaneous SNR received at R_1 can be expressed as [25]

$$F_{\gamma_{S,R_1}}\left(\gamma_{S,R_1}\right) = 1 - Q_1\left[\sqrt{2K_1}, \sqrt{\frac{2\left(K_1+1\right)}{\overline{\gamma}_{S,R_1}}\gamma_{S,R_1}}\right], \quad (21)$$

where $Q_1[\cdot]$ is the first-order Marcum Q-function. The CDF $F_{\gamma_{R_2,D}}\left(\gamma_{R_2,D}\right)$ of the Rician distribution for the second RF link can be similarly defined by replacing the indices.

The CDF of the instantaneous SNR received at R_2 can be obtained as follows

$$F_{\gamma_{R_1,R_2}}\left(\gamma_{R_1,R_2}\right)=\int_0^{\gamma_{R_1,R_2}} f_{\gamma_{R_1,R_2}}(x)dx. \quad (22)$$

Using (19) and [26, eq. (07.34.21.0003.01)], (22) becomes

$$F_{\gamma_{R_1,R_2}}\left(\gamma_{R_1,R_2}\right)=\frac{ab\gamma^2}{AI^l\Gamma(a)\Gamma(b)}\sqrt{\frac{\gamma_{R_1,R_2}}{\overline{\gamma}_{R_1,R_2}}}$$
$$\times G_{2,4}^{3,1}\left[\frac{ab}{AI^l}\sqrt{\frac{\gamma_{R_1,R_2}}{\overline{\gamma}_{R_1,R_2}}}\middle|\begin{matrix}0,\gamma^2\\ \gamma^2-1,a-1,b-1,-1\end{matrix}\right]. \quad (23)$$

The expressions in (21) and (23) can be numerically evaluated using well-known software packages, e.g., Mathematica and MATLAB.

V. RESULTS

In this section, the performance of the proposed system is evaluated in terms of the outage probability. Unless indicated otherwise, the values of model parameters used are $\overline{\gamma}_{S,R_1}=\overline{\gamma}_{R_2,D}=20$ dB, $\overline{\gamma}_{R_1,R_2}=50$ dB, $\gamma_{out}=0$ dB, K_1 = K_2 = K = 10 dB, L = 400 km, h = 20 km, σ = 0.01 dB/km (clear weather conditions), $w=w_0=2$ cm, Θ_0 = 0 (focus Gaussian beam), and $\sigma_s=0.15$ m. Note that a laser aperture diameter of $2a_r=0.3$ m is assumed. Then, $w_z=1.38$ m [27]. A wavelength λ = 1550 nm is considered for the FSO link since it is widely used for most airborne optical communication scenarios. In addition, two values of the wind speed that determine C_n^2 are considered; $u_1=10$ m/s (best case) and $u_2=30$ m/s (worst case) [23]. For these values, we obtain $C_{n,u_1}^2\approx1.72\cdot10^{-19}$ m$^{-2/3}$, $C_{n,u_2}^2\approx1.55\cdot10^{-18}$ m$^{-2/3}$, $\sigma_{R,u_1}^2\approx0.2$, $\sigma_{R,u_2}^2\approx1.82$, $\sigma_{I,u_1}^2\approx$ 0.08, and $\sigma_{I,u_2}^2\approx0.67$.

Figure 2 depicts the outage probability for different values of the Rician factor K_1 = K_2 = K and u = 10 m/s. It is obvious that increasing the Rician factor improves the performance of the system. Note that a certain outage floor exists at high values of K due to the average SNR over the FSO link. Hence, all curves tend to the same saturation floor.

Figure 3 demonstrates the outage probability for both the best and worst case of turbulence conditions and different inter-HAP distance. As this distance decreases, a significant performance improvement can be achieved. However, the variation of the FSO propagation distance slightly affects the outage probability when the worst case of turbulence conditions is observed.

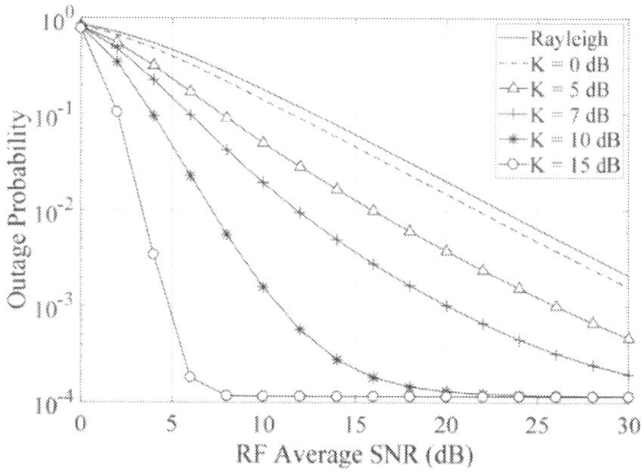

Figure 2. The outage probability in terms of the RF average SNR for different Rician factor of the RF links.

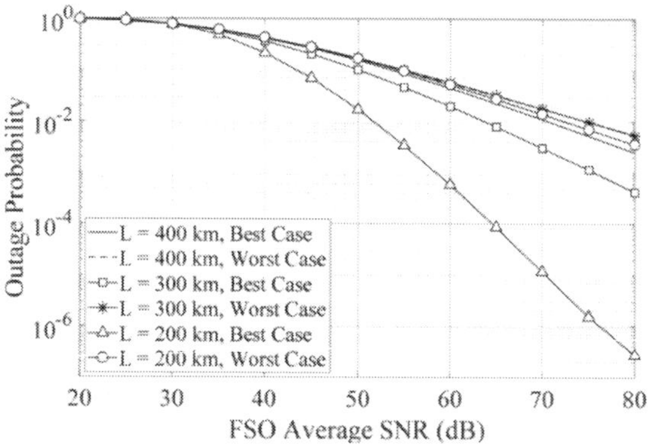

Figure 3. The outage probability in terms of the FSO average SNR for different turbulence conditions and different inter-HAP distance.

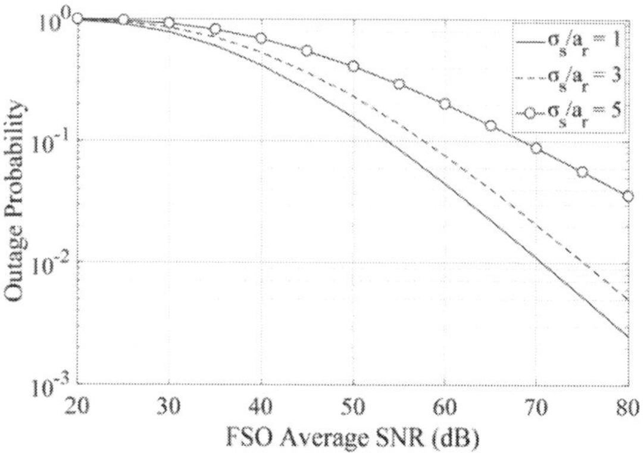

Figure 4. The outage probability in terms of the FSO average SNR for different normalized jitter standard deviation.

Finally, Figure 4 shows the effect of the normalized jitter standard deviation σ_s / a_r on the outage probability for $u = 10$ m/s. One observes that the performance significantly degrades and the pointing errors have a dominant effect on the system performance, as σ_s / a_r increases.

VI. CONCLUSION

In this paper, the performance of a triple-hop mixed RF/FSO/RF stratospheric communication system has been investigated. The results have depicted that the Rician factor significantly influences the system performance. These results have also underlined that the performance degrades, as the inter-HAP distance increases. However, for the worst turbulence case, this distance slightly affects the outage probability. Finally, the results have shown that the misalignment-induced fading has a severe effect on the system performance. Since there are no experimental data available in the literature to fully verify the theoretical results, future research efforts may be devoted to collecting measured channel data in real-world propagation conditions.

ACKNOWLEDGMENT

This research is implemented through State Scholarships Foundation (IKY) and co-financed by the European Union (European Social Fund – ESF) and Greek national funds through the action entitled "Reinforcement of Postdoctoral Researchers" in the framework of the Operational Programme "Human Resources Development Program, Education and Lifelong Learning", with priority axis 6, 8, 9, of the National Strategic Reference Framework (NSRF) 2014-2020.

REFERENCES

[1] R. Giuliano, M. Luglio, and F. Mazzenga, "Interoperability between WiMAX and broadband mobile space networks," *IEEE Commun. Mag.*, vol. 46, no. 3, pp. 50-57, Mar. 2008.

[2] A. Mohammed, A. Mehmood, F. Pavlidou, and M. Mohorcic, "The role of high-altitude platforms (HAPs) in the global wireless connectivity," *Proc. IEEE*, vol. 99, no. 11, pp. 1939-1953, Nov. 2011.

[3] B. Moision et al. "Demonstration of free-space optical communication for long-range data links between balloons on Project Loon," in *Proc. SPIE 10096*, Feb. 2017.

[4] H. Kaushal and G. Kaddoum, "Optical Communication in Space: Challenges and Mitigation Techniques," *IEEE Comm. Surv. & Tutorials*, vol. 19, no. 1, pp. 57-96, Firstquarter 2017.

[5] D. Giggenbach, R. Purvinskis, M. Werner, and M. Holzbock, "Stratospheric Optical Inter-Platform Links for High Altitude Platforms," in *Proc. 20th AIAA ICSSC*, Montreal, pp. 1-11, May 2002.

[6] L. Andrews, R. L. Philips, and C. Y. Hopen, *Laser Beam Scintillation With Applications*. SPIE Press, 2001.

[7] M. Sharma, D. Chadha, and V. Chandra, "High-altitude platform for free-space optical communication: Performance evaluation and reliability analysis," *J. Opt. Commun. Netw.*, vol. 8, no. 8, pp. 600-609, Aug. 2016.

[8] C. Datsikas, K. P. Peppas, N. C. Sagias, and G. S. Tombras, "Serial free-space optical relaying communications over gamma-gamma atmospheric turbulence channels," *J. Opt. Commun. Netw.*, vol. 2, pp. 576-586, Jul. 2010.

[9] A. Aragón-Zavala, J. L. Cuevas-Ruiz, and J. A. Delgado-Penín, *High Altitude Platforms for Wireless Communications*, 1st ed. New York: Wiley, Dec. 2008.

[10] F. S. Vetelino, C. Young, L. Andrews, and J. Recolons, "Aperture averaging effects on the probability density of irradiance fluctuations in moderate-to-strong turbulence," *Applied Optics*, vol. 46, no. 11, pp. 2099-2108, Apr. 2007.

[11] A. M. Salhab, "A New Scenario of Triple-Hop Mixed RF/FSO/RF Relay Network with Generalized Order User Scheduling and Power Allocation," *EURASIP Journal on Wireless Communications and Networking* (2016) 2016:260.

[12] E. Lee, J. Park, D. Han, and G. Yoon, "Performance analysis of the asymmetric dual-hop relay transmission with mixed RF/FSO links," *IEEE Photonic Tech. L.*, vol. 23, no. 21, pp. 1642-1644, 2011.

[13] S. M. Navidpour, M. Uysal, and M. Kavehrad, "BER performance of free-space optical transmission with spatial diversity," *IEEE Trans. Wirel. Commun.*, vol. 6, no. 8, pp. 2813-2819, Aug. 2007.

[14] M. K. Simon and M.-S. Alouini, *Digital Communication over Fading Channels*, 2nd ed. New York: Wiley, 2005.

[15] Iskandar and S. Shimamoto, "Channel characterization and performance evaluation of mobile communication employing stratospheric platforms," *IEICE Trans. on Commun.*, vol. E-B, no. 3, pp. 937-944, Mar. 2006.

[16] X. Zhu and J. Kahn, "Free space optical communication through atmospheric turbulence channels," *IEEE Trans. Commun.*, vol. 50, no. 8, pp. 1293–1300, Aug. 2002.

[17] H. G. Sandalidis, "Coded free-space optical links over strong turbulence and misalignment fading channels," *IEEE Trans. Commun.*, vol. 59, no. 3, pp. 669-674, 2011.

[18] A. A. Farid, S. Member, and S. Hranilovic, "Outage capacity optimization for free-space optical links with pointing errors," *J. Lightwave Technol.*, vol. 25, pp. 1702-1710, Jul. 2007.

[19] M. A. Al-Habash, "Mathematical model for the irradiance probability density function of a laser beam propagating through turbulent media," *J. Opt. Eng.*, vol. 40, no. 8, pp. 1554-1562, Aug. 2001.

[20] V. S. Adamchik and O. I. Marichev, "The algorithm for calculating integrals of hypergeometric type functions and its realization in REDUCE system," in *Proc. Int. Conf. on Symbol. and Algeb. Comp.*, Tokyo, Japan, pp. 212-224, 1990.

[21] Y. Ren, A. Dang, B. Luo, and H. Guo, "Capacities for long-distance free-space optical links under beam wander effects," *IEEE Photon. Techn. Lett.*, vol. 22, no. 14, pp. 1069-1071, Jul. 2010.

[22] L. Andrews and R. L. Philips, *Laser Beam Propagation Through Random Media*, 2nd ed. Bellingham, WA: SPIE, pp. 350-351, 2005.

[23] S. Parthasarathy, D. Giggenbach, and A. Kirstädter, "Channel modelling for free-space optical inter-HAP links using adaptive ARQ transmission," in *Proc. SPIE 9248*, 92480Q, pp. 1-11, Oct. 2014.

[24] S. S. Ikki and S. Aissa, "A study of optimization problem for amplify-and-forward relaying over Weibull fading channels with multiple antennas," *IEEE Commun. Lett.*, vol. 15, no. 11, pp. 1148-1151, 2011.

[25] A. A. Abu-Dayya and N. C. Beaulieu, "Switched diversity on microcellular Ricean channels", *IEEE Trans. Veh. Technol.*, vol. 43, no. 4, pp. 970-976, 1994.

[26] The Wolfram Functions Site, 2013. [Online]. Available from: http://functions.wolfram.com [retrieved: March, 2018].

[27] N. Vaiopoulos, H. G. Sandalidis, and D. Varoutas, "Using a HAP network to transfer WiMAX OFDM signals: Outage probability analysis," *J. Opt. Commun. Netw.*, vol. 5, pp. 711-721, Jul. 2013.

Generation and Propagation Characteristics Analysis through Atmosphere of Gaussian-Bessel-correlated Schell-model Beam

Zeyu Zhou[1], Xiuhua Yuan[*1], Wei Yao[2]
1: School of Optical and Electronic Information, Huazhong University of Science and Technology, Wuhan China
2: Shenzhen Research Institution, Huazhong University of Science and Technology, Shenzhen, China
Email: *yuanxh@hust.edu.cn, zhouzeyu@hust.edu.cn, yaowei@163.com

Abstract—**In this paper, propagation characteristics of Gaussian-Bessel-correlated Schell-model beam (GBSM) in turbulent atmosphere have been studied through simulation and experimentation. Without studying its quality in communication directly, we use scintillation index (SI) and beam expansion to estimate such beams as signal lights. We use Monte-Carlo method to generate the phase screen for modeling the atmosphere in simulation. An experiment consisting of a phase-only spatial light modulator (SLM) and a rotating ground-glass disk (RGGD) is used to generate a GBSM beam. A box with fans and heaters inside creates a real, but much stronger, turbulent atmosphere environment along the propagation path. Comparing to other kinds of beams, like conventional Gaussian Schell-model beam (GSM), GBSM has a faster speed of beam expansion and better fluctuation characteristics in far propagation range while GSM and GBSM have the same initial coherence radius.**

Keywords- GBSM; generation; turbulence; propagation.

I. INTRODUCTION

Since the appearance of the laser, it has been used in fields of communication, detection, illumination, imaging and even in military for its advantages in coherence and orientation. However, influenced by atmospheric turbulence, especially near the ground, the front wave of the beam breaks and the beam intensity fluctuates, which limits the usage of laser in atmosphere. Since Ronald Fante [1] proved in 1981 that a partially coherent beam has extreme improvement on the intensity fluctuation, people have studied many kinds of beams with different coherence profiles. Besides the conventional Gaussian Schell-model beams (GSB) [2], beams with non-Gaussian coherence degree like pseudo-Bessel-correlated beams (PBCB) [3], multiple Gaussian Schell-model beams (MGSB) [4], Laguerre-Gaussian Schell-model beams (LGSB) [5], rectangular Gaussian Schell-model beams [6] have also been studied in turbulence.

All of such beams with spatial partial coherence are better than coherent Gaussian beam with respect to the scintillation index. Moreover, the lower the coherence degree, the lower the SI. Since so many achievements have come out in this field, scintillation of GBSM still remains its value for study. Thus, we want to further compensate this work.

The rest of the paper is structured as follows. In Section 2, a method to generate GBSM is introduced in theory and tested in an experiment. Afterwards, the average intensity distribution of GBSM beam is deducted theoretically, and the intensity fluctuation quality is analyzed via simulation in Section 3. We conclude the paper in Section 4.

II. GENERATION OF GBSM

A. Gaussian-Bessel-correlated Schell-model beam

The GBSM is a partially coherent beam with non-conventional correlation form defined by Gaussian-Bessel function. Even though it has stochastic field, the GBSM can be described by auto-correlation function, a second-order moment of optical field $U(r)$, as (1)

$$\langle U(\vec{r_1})U^*(\vec{r_2})\rangle = \langle U(\vec{r_1})\rangle\langle U^*(\vec{r_2})\rangle J_0\left(\frac{|\vec{r_1}-\vec{r_2}|}{\alpha}\right) exp\left(-\frac{|\vec{r_1}-\vec{r_2}|^2}{\beta^2}\right), \tag{1}$$

where the superscript asterisk means conjugation of the field, and the angular bracket means ensemble average. J_0 here means the first kind of Bessel function on zero-order. α is the parameter controlling the Bessel component, and β determines the truncation width of exponential function.

To generate such a beam, we use a Kepler telescope structure with a RGGD [7] placed on the common focus of two lenses, see Figure 1. When a Gaussian-Bessel beam,

$$U(\vec{s}) = J_0\left(\frac{|\vec{s}|}{\alpha_0}\right) exp\left(-\frac{s^2}{\beta_0^2}\right) \tag{2}$$

where $\vec{s}$ is a vector which describe the location of arbitrary field point on plane in front of the first lens f_1, illuminates the first lens of the telescope, there will appear an annular beam spot on the RGGD. Assuming the field through the RGGD becomes totally incoherent, we can consider the field behind RGGD as a random field which obeys δ distribution,

$$M(\vec{t_1},\vec{t_2}) = \langle U(\vec{t_1})U^*(\vec{t_2})\rangle = \sqrt{I(\vec{t_1})I(\vec{t_2})}\,\delta(\vec{t_1},\vec{t_2}) \tag{3}$$

where $I(t)$ means the beam intensity on point t,

$$I(\vec{t}) = U(\vec{t})U^*(\vec{t}). \tag{4}$$

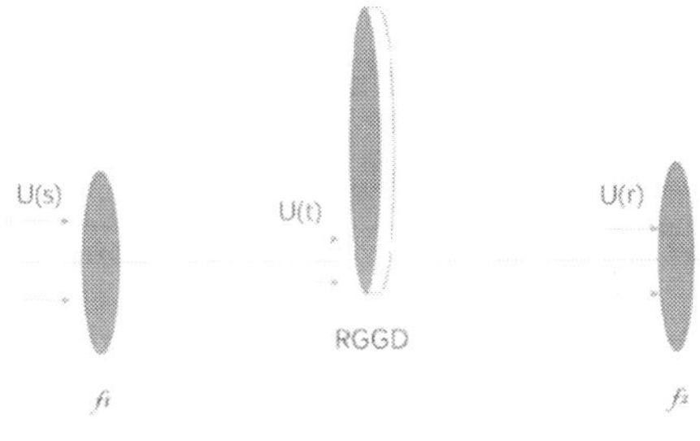

Figure 1. GBSM beam generation illustration

With Fresnel diffraction theory, we can find optical field *U(t)* in (4) is Fourier transformation of *U(s)* in (2). It has the convolution between an annular impulse and a Gaussian function

$$\mathrm{U}(\vec{t}) \propto exp\left(\frac{ik}{2f_1}t^2\right)\delta\left(|\vec{t}| - \frac{\lambda f_1}{2\pi\alpha_0}\right) * exp\left(-\frac{\pi^2\beta_0^2 t^2}{\lambda^2 f_1^2}\right). \quad (5)$$

When α is far smaller than β, we can consider that *I(t)* becomes a convolution between square of two parts as (6)

$$\mathrm{I}(\vec{t}) \propto \delta\left(|\vec{t}| - \frac{\lambda f_1}{2\pi\alpha_0}\right) * exp\left(-\frac{2\pi^2\beta_0^2 t^2}{\lambda^2 f_1^2}\right). \quad (6)$$

After passing through the second lens f_2, the auto-correlation function of *U(r)* can be obtained,

$$\begin{aligned}\langle U(\vec{r_1})U^*(\vec{r_2})\rangle &= \left(\frac{k}{2\pi f_2}\right)^2 \iint_{-\infty}^{\infty} d^2t_1 \iint_{-\infty}^{\infty} d^2t_2 \mathrm{M}(\vec{t_1},\vec{t_2}) \\ &\times exp\left[\frac{ik}{2f_2}|\vec{t_1}-\vec{r_1}|^2 - \frac{ik}{2f_2}|\vec{t_2}-\vec{r_2}|^2\right] exp\left(-\frac{ik}{2f_2}r_1^2 + \frac{ik}{2f_2}r_2^2\right) \\ &= \left(\frac{k}{2\pi f_2}\right)^2 \iint_{-\infty}^{\infty} d^2t \mathrm{I}(\vec{t}) exp\left[\frac{ik}{f_2}(\vec{r_1}-\vec{r_2})\cdot\vec{t}\right] \\ &\propto J_0\left(\frac{f_1|\vec{r_1}-\vec{r_2}|}{f_2\alpha_0}\right) exp\left(-\frac{f_1^2|\vec{r_1}-\vec{r_2}|^2}{2f_2^2\beta_0^2}\right)\end{aligned} \quad (7)$$

In the deduction of (7), considering the integration in the third line is, mathematically, a Fourier transformation of *I(t)* which is represented in (6), it is convenient to derive to the last result by multiplying the Fourier transformations of each part in convolution of *I(t)*. By changing the parameters of the incident beam, we can adjust the coherence quality conveniently.

B. Generation of GBSM in experiment

From Figure 1, we establish an experiment structure to generate a GBSM beam, which is presented in Figure 2. We use He-Ne laser (632.8nm) as the laser source. Phase-only SLM (HOLOEYE PLUTO) is used instead of an axicon to generate a Gaussian-Bessel beam. After passing through the telescope structure with RGGD, a partially coherent beam comes out. As the glass is using 320 mesh quartz sand to grind, the RGGD has a coherent length of about 80um, making every 1mm^2 illumination area include about 150 coherent units, which can be regarded to meet the demand for δ distribution. The RGGD has a diameter of 60mm, when the optical axis is 20mm from the disk center, and illumination area about 2mm^2, rotation speed adjusted to

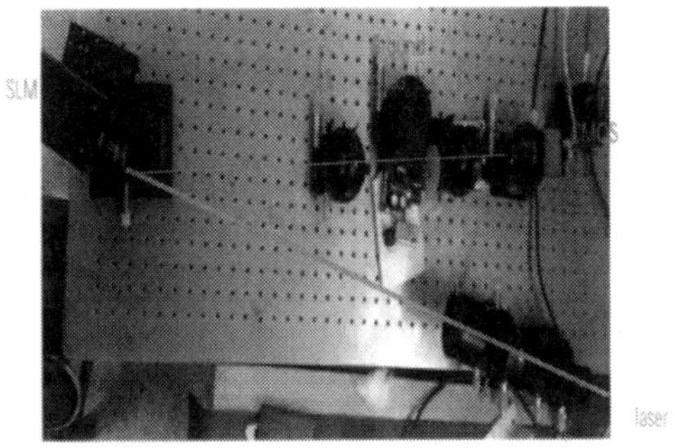

Figure 2. Experiment illustration of GBSM generation. SLM: spatial light modulator. f_1, f_2: lenses with focus length of f_1 and f_2 respectively. Laser: 632.8nm He-Ne laser

1r/s, the coherent time of the beam will be fixed at about 50us. To observe this beam, we use a CMOS with exposure time of 200us. The beam in CMOS is presented in Figure 3(a). According to Gaussian momentum theory [8], we calculate the coherence degree of optical field through count the intensity variation from 2000 pictures and finally print it in Figure 3(b). Even though the graph is not excellently matching the expectation, we can still tell the tendency of Gaussian-Bessel function. Many factors lead to this deviation. The Bessel-Gaussian beam created after SLM is not accurate enough. Crosstalk between the fringes exists. Furthermore, coherence length of RGGD is still too large to

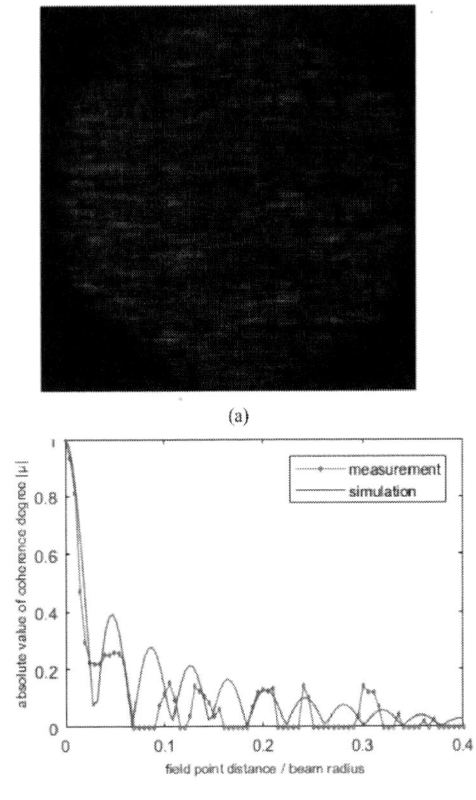

(b)

Figure 3. Generation of GBSM, (a) is the picture of a sample from CMOS and (b) is the curve graph of coherence degree varying with the field point distance relative to beam radius

form the δ distribution, which makes the coherence degree out of its shape. Also, because of the improper set of axis, we miss the details and since the interval between each pixel on the CMOS is limited, it is unavailable for us to adjust the axis setting to a proper one.

III. PROPAGATION CHARACTERISTICS OF GBSM

A. *Average intensity of GBSM in atmospheric turbulence*

The extended Huygens-Fresnel theory has the advantage to calculate such complex random optic field. As compared to the Rytov approximation, this method avoids to deduce the first and second order perturbation from turbulence to optic field, which is difficult and worthless. This theory divides the entire field into countless spherical secondary sources, and the influence from randomly varying refractive on the wave front can be described by the widely acceptable complex phase perturbation factor for spherical wave.

The atmospheric turbulence is described by turbulence eddies and turbulence influences the beam propagation by randomly changing reflective index. We deem that the reflective index *n(R)* has a spatial power spectrum

$$\Phi_n(\kappa) = 0.033C_n^2\kappa^{-11/3}exp\left(-\frac{\kappa^2}{\kappa_l^2}\right) \tag{8}$$

which is named Tatarskii spectrum. C_n^2 is refractive index structure constant used to describe the fluctuation intensity of turbulence. $\kappa_l = \frac{5.92}{l_0}$ represents the spatial frequency of inner scale l_0. Utilizing extended Huygens-Fresnel theory, the average intensity of beam after propagating a length of L in atmospheric turbulence can be written as

$$\mathrm{I}(\vec{\rho}, L) = \left(\frac{k}{2\pi L}\right)^2 \iint \iint_{-\infty}^{\infty} d^2r_1 d^2r_2 \langle U(\vec{r_1}, 0)U^*(\vec{r_2}, 0)\rangle$$
$$\times\, exp\left[\frac{ik}{2L}|\vec{r_1} - \vec{\rho}|^2 - \frac{ik}{2L}|\vec{r_2} - \vec{\rho}|^2\right]\langle exp[\phi(\vec{\rho}, \vec{r_1}) + \phi^*(\vec{\rho}, \vec{r_2})]\rangle \tag{9}$$

where ϕ stands for the additional phase from the turbulence eddies through which the beam passes. We have assumed the stochastic behavior of optic source and that of refractive distribution are stochastically independent. The part in angular bracket can be written as [9]

$$\langle exp[\phi(\vec{\rho}, \vec{r_1}) + \phi^*(\vec{\rho}, \vec{r_2})]\rangle \approx -0.145C_n^2k^2L\kappa_l^{1/3}Q^2 \tag{10}$$

where we use an algebraic substitution

$$\vec{Q} = \vec{r_1} - \vec{r_2}. \tag{11}$$

If we also make

$$\vec{R} = \frac{1}{2}(\vec{r_1} + \vec{r_2}) \tag{12}$$

and substitute (1) and (10-11) into (9), after integrating, the average intensity turn into a convolution

$$\mathrm{I}(\vec{\rho}, L) \propto \delta(|\vec{\rho}| - T) * exp\left(-\frac{\rho^2}{w^2}\right), \tag{13}$$

where

$$\mathrm{T} = \frac{L}{2\pi\alpha k} \tag{14}$$

is the radius of annular beam profile, and

$$w^2 = \frac{2\pi^2L^2}{k^2w_0^2} + \frac{\pi^2w_0^2}{2} + \frac{4\pi^2L^2}{k^2\beta^2} + 5.724C_n^2k^2L^3\kappa_l^{1/3} \tag{15}$$

is the beam expansion which result from free space diffraction and turbulence fluctuation. Through (14) and (15), we can see the radius of annular beam profile has nothing to do with turbulence fluctuation. Apart from propagation length and wavelength, only the parameter of Bessel function part has the ability to determine the ring's radius. Since the inner scale l_0 is in the magnitude of millimeter, when it is in short propagation length, the second part of (15) is the majority determines the beam radius, instead of the part associated with β. Hence, the ring profile cannot be observed if T<w, or

$$\alpha > \frac{L}{\sqrt{2\pi^2}kw_0}. \tag{16}$$

Making α reach to infinity, Eq. (13) reduces to the intensity distribution of GSM. Beam expansion (15) becomes beam radius

$$w_G^2 = \frac{2\pi^2L^2}{k^2w_0^2} + \frac{\pi^2w_0^2}{2} + \frac{4\pi^2L^2}{k^2\beta_G^2} + 5.724C_n^2k^2L^3\kappa_l^{1/3}. \tag{17}$$

Compared to GSM, the GBSM beam, with a same coherence length, obviously has a larger beam expansion. However, if the hollow profile becomes useful, the GBSM beam may have advantage over GSM.

B. *Scintillation index*

Even though we have given out the average intensity of beam after propagating through turbulence, we have no idea how the intensity fluctuation is. The Monte-Carlo method helps us create a phase screen through which the optical beam will get a randomly changing phase factor. This phase factor represents the influence of turbulence on the laser beam. In order to reduce the simulation error, the propagation of laser beam is devised into several parts. For each part, the angular spectrum method is used to calculate the output field.

The scintillation index is a normalized variance of intensity, which is used to estimate the intensity fluctuation of the laser beam. SI is defined as below,

$$\sigma_I^2 = \frac{\langle I^2\rangle}{\langle I\rangle^2} - 1. \tag{18}$$

For a communication application, a low intensity fluctuation is of much expectation. Because the receiver always has an effective area, leading to a so called aperture averaging, analyzing the total energy fluctuation on a fixed area is more helpful. And, since both the optical field and the channel media have random characteristics, a large number of samples are necessary. After about 2000 realizations, the statistical result becomes stable.

First, we check the impact of Bessel parameter on the scintillation index. We make α vary from 0.002m to infinite at which the light becomes GSM. Gaussian parameter β is fixed on the value of 0.05m. Detection diameter is set to 50mm. Their scintillation indices on the observation plane, which is placed several different distances away from transmission plane, are presented in Figure 4(a). Same as our recognition of relationship between scintillation and beam coherence degree, the SI decreases with the beam coherence length which, in this case, is determined by α, except the infinite one by β. Figure 4(b) is the responding beam radius variation, which is contrary to SI.

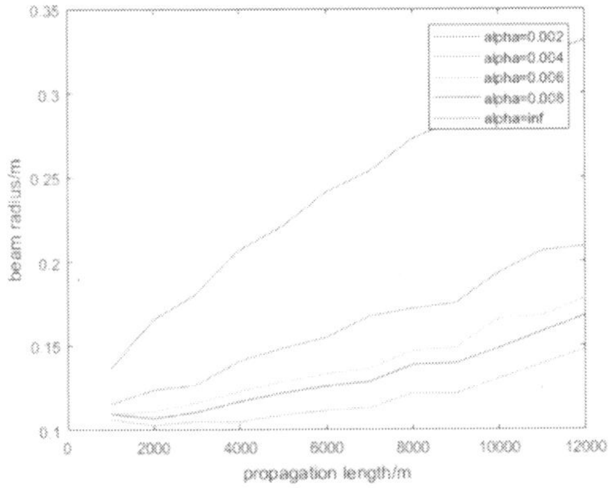

(a)

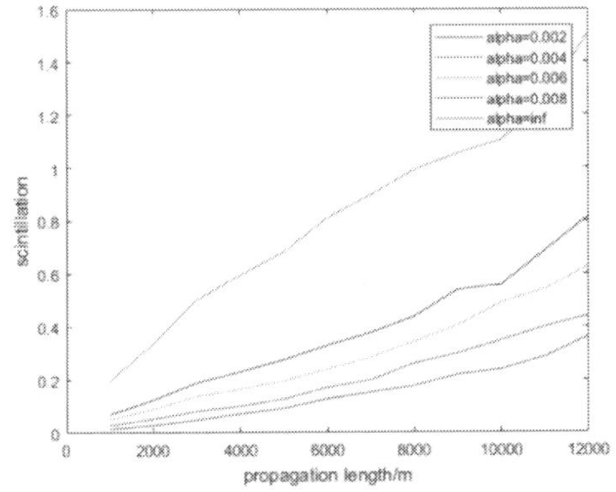

(b)

Figure 4. Scintillation indices (a) and beam radius (b) of GBSM beams with different Bessel parameter α of 0.002, 0.004, 0.006, 0.008, infinite varying with propagation length

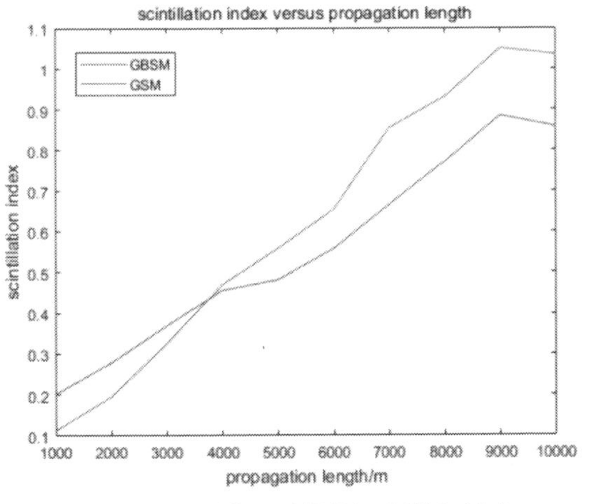

Figure 5. Scintillation indices of GBSM and GSM with the same coherence length of 3.5mm versus propagation length

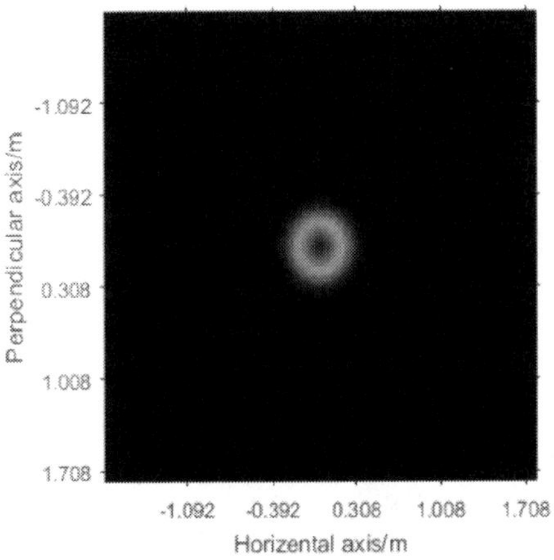

(a)

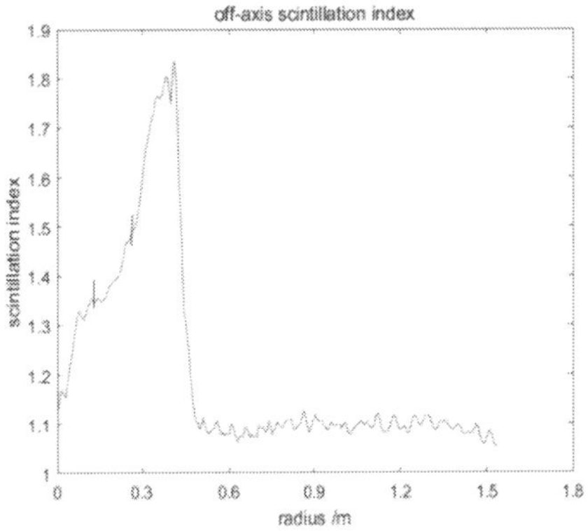

(b)

Figure 6. Intensity distribution (a) and off-axis scintillation (b) of GBSM with Bessel parameter of 0.001 after propagating 6km in turbulence

Even though we have studied the effect of coherence parameter, the profile of coherence degree distribution is still of much importance. As conventional GSM beam has already proved to be better than coherent Gaussian beam, we use GSM only to compare with GBSM beam. In this simulation, the laser wavelength is 1.55 um. Coherence length and beam width is fixed to about 3.5 mm and 14 cm, respectively. The result of the scintillation index on the detection area of such two kinds of beams varying with propagation length is presented in Figure 5. According to the result, obviously, the GBSM beam has a larger on-axis scintillation in a short range, but it becomes the smaller one in the long range. The reason of this result is unclear. There is only a surmise of it, which is based on the evolution of intensity distribution of GBSM. As the ring-like profile this beam has in a short range, even it may be blurry when the ring's radius is less than the Gaussian expansion, the lower level of average energy distribution on the center point leads to a larger relative fluctuation. With the further propagation of the beam, the ring of GBSM diffuses, but slowly comparing to GSM, which making more energy left in the center area. This may control the scintillation increasing slowly.

In the last part, we determined the optical intensity distribution. We saw the beam profile would become a hollow. It is of considerable need to research the off-axis SI. According to (16), setting α to be 0.001m and β still 0.05m, after 6km propagation, the beam profile and off-axis scintillation index is presented in Figure 6. The interval in Figure 6(b) between two vertical lines corresponds to the ring part of profile in Figure 6(a). There are two rapid rises separated by the ring area. Without theoretical analysis, the reason why it happens is unclear.

IV. CONCLUSION

In this paper, we have analyzed the average intensity and scintillation index of GBSM. Compared to GSM, it has an annular intensity profile, leading a larger beam intensity distribution. However, the radius of the annular profile is fixed by Bessel parameter and propagation length, making the beam somewhat controllable and practical. Its scintillation index is lower over a long distance, but rapid rise of scintillation index off the axis likely causes the application limitation of GBSM. We have generated a GBSM beam using a SLM and an RGGD. The result seems acceptable.

Next, we plan to use this beam as a communication light source, testing its application in communication. Since the partial coherence of beam leads to intensity fluctuation even in initial plan of beam propagation, but there is a time average process when the beam is used in communication, so it is necessary to analyze the influence from coherence time of the source, observer and random media on scintillation index.

ACKNOWLEDGMENT

This research is supported by Shenzhen Research Institution of Huazhong University of Science and Technology.

REFERENCES

[1] R. L. Fante, "Intensity Fluctuations of an Optical Wave in a Turbulent Medium Effect of Source Coherence," Optica Acta. vol. 28, no. 9, pp. 1203-1207, 1981.

[2] G. Gbur and E. Wolf, "Spreading of partially coherent beams in random media," J. Opt. Soc. Am. A, vol. 19, no. 8, pp. 1592-1598, 2002.

[3] Y. Gu and G. Gbur, "Scintillation of pseudo-Bessel correlated beams in atmospheric turbulence," J. Opt. Soc. Am. A, vol. 27, no. 12, pp. 2621-2629, 2010.

[4] Y. Yuan and Y. Cai, "Scintillation index of a multi-Gaussian Schell-model beam in turbulent atmosphere," Optics Communications, vol. 305, pp. 57-65, 2013.

[5] R. Chen et al, "Statistical properties of a Laguerre-Gaussian Schell-model beam in turbulent atmosphere," Optics Express, vol. 22, no. 2 pp. 1871-1883, 2014.

[6] O. Korotkova and E. Shchepakina, "Rectangular Multi-Gaussian Schell-Model beams in atmospheric turbulence," Journal of Optics, vol. 16, no. 4, pp. 1-8, 2014.

[7] Y. Cai, Y. Chen and F. Wang, "Generation and propagation of partially coherent beams with nonconventional correlation functions: a review [Invited]," J. Opt. Soc. Am. A, vol. 31, no. 9, pp. 2083-2096, 2014.

[8] L. Mandel and E. Wolf, "Coherence and Quantum Optics," Cambridge: Cambridge University Press, 2001.

[9] L. C. Andrews and R. L. Phillips, "Laser beam propagation through random medium," SPIE Press, 1998.

Application of Aperture Truncated Airy Beams in Free Space Optical Communications

Minghao Wang[1,2], Xiuhua Yuan[1], Peng Deng[2], Wei Yao[3], Timothy Kane[2]
[1]School of Optical and Electronic Information, Huazhong University of Science and Technology, Wuhan, China
[2]Department of Electrical Engineering, The Pennsylvania State University, University Park, USA
[3]Shenzhen Research Institute of Huazhong University of Science and Technology, Shenzhen, China.
e-mails: wmh@hust.edu.cn, yuanxh@hust.edu.cn, pxd18@psu.edu, weiyao@hust.edu.cn, tjk7@psu.edu

Abstract—**Free Space Optical Communications (FSOC) has proved to be a reliable alternative to radio frequency systems in providing high speed wireless data links, especially for satellite communications. To further improve the FSOC performance, plenty of novel light source beams are proposed and evaluated, among which Airy beams have gained intensive research efforts in the past decade, owing to its remarkable features. An ideal Airy beam is able to propagate in free space without diffraction, and it is resistant to atmospheric turbulence due to the self-healing feature. Besides, the main lobe of an Airy beam follows a controllable ballistic trajectory, which can be particularly useful to improve the link robustness in presence of blockage. However, if the transverse extent of the Airy beam is confined by a finite optical aperture, the non-diffractive features no longer retain. But within the limited diffraction-free range, the Airy beam is still able to self-heal and self-bend, making it valuable for enhancing the survivability of FSOC systems. In this study, we first examined the effects of truncation methods on the Airy diffraction-free range and found out that the aperture-truncated Airy (ati-Airy) beam is more preferable in terms of propagation distance. Then, the scintillation behavior of ati-Airy beams is investigated under the conditions of controlled self-bending as well as self-healing from obstructions.**

Keywords-free space optical communications; Airy beam; truncated Airy beams; beam propagation; scintillation index.

I. Introduction

With the ever increasing demand of high bandwidth in satellite data transmission, in recent years Free Space Optical Communications (FSOC) has become one of the mainstream physical-layer technologies [1][2]. Most of the current FSOC systems use traditional Gaussian beams as the optical source. In order to improve the receiving performance, various novel beams have been proposed [3]-[6], among which is the Airy beam [7]. The reason why the Airy beam is considered promising in FSOC roots in its remarkable features—non-diffracting propagation in free space [8], self-accelerating in the transverse direction [9] and self-healing in the presence of obstacles [10].

In the context of FSOC, it is of great beneficial if optical beams can propagate without diffraction, since the power loss due to diffractive spreading can be minimized in long-haul links. However, for an Airy beam to fulfill the ideal diffraction-free feature, the transmitter (TX) optics should be larger than the spatial extent of the Airy field, which is determined by $k^2x_0^3$ [11], where k is the optical wave number and x_0 is the characteristic length of the Airy function. For effective capture and tracking in a practical system, x_0 should not be too small, and reasonable values of x_0 would require intimidatingly large TX apertures. That being said, in the real world of FSOC, the transmitting beam is always truncated by the finite TX aperture. Fortunately when that happens, the Airy beam is still able to propagate without much spreading within a limited distance called (nearly) diffraction-free range, beyond which severe diffraction takes place.

Despite its limited propagation distance, the truncated Airy beam is appealing. The well preserved self-accelerating feature makes it unique as an optical carrier, bringing the possibility of circumventing obstacles on the path to the receiver. If the obstacle is too big to circumvent, the Airy beam can even regenerate the blocked part of its optical field, which is known as self-healing. Therefore, optical links built with Airy beams can be more robust than the existing line-of-sight (LOS) links, especially in satellite communication scenarios, where blockage caused by clouds or other objects frequently occurs.

In this study we try to demonstrate the performance of truncated Airy beams in FSOC links. By means of wave optics simulation (WOS), the effects of various truncation methods on the diffraction-free range are numerically examined. Reaching at the conclusion that the aperture truncated ideal Airy beam is diffracted least, we further investigate its scintillation behavior when the ballistic trajectory is dynamically controlled. Then, by blocking the main lobe at the initial plane or at the halfway point of propagation, we have managed to analyze the aperture-averaged scintillation, as well as the mean Signal-to-Noise Ratio (SNR) of the regenerated main lobe. In addition to horizontal links, the propagation through vertically distributed turbulent atmosphere is also studied with non-uniformly distributed phase screens.

The rest of this paper is structured as follows. Section II is intended to find out in the presence of a circular hard aperture, which truncation method enables the longest diffraction-free propagation distance. Then, in Section III, the self-accelerating and self-healing properties of truncated Airy beams are examined from the perspective of scintillation index. Section IV is a brief summary of the presented work.

II. DIFFRACTION FREE RANGE OF TRUNCATED AIRY BEAMS

The Airy functions are the solution of the paraxial wave equation of diffraction, and for (1 + 1) D case the beam wave envelope is given as

$$U(s,\xi) = Ai(s-\xi^2/4)\exp\left[i(6s\xi-\xi^3)/12\right], \quad (1)$$

where $s = x/x_0$ represents the transverse scale normalized by the Airy characteristic length x_0, $\xi = z/kx_0^2$ is the normalized propagation distance, and $k = 2\pi/\lambda$ is the optical wavenumber.

In order to dynamically control the drop point, or the trajectory of the main lobe, the "initial velocity" parameter υ is introduced into the initial plane expression which leads to

$$U(s,0) = Ai(s)\exp\left[i\upsilon s\right], \quad (2)$$

from which the transverse shift of the main lobe s_0 can be determined by

$$s_0 = \xi^2/4 + \upsilon\xi. \quad (3)$$

Then, we can see that the acceleration of the ballistic is actually immutable (in the normalized s-ξ coordinate it is 1/2), yet the drop point of the main lobe can be easily controlled via υ at a specified distance ξ.

Let us rewrite a 2D ideal Airy beam in the initial plane in this way:

$$U(s_x,s_y) = Ai(s_x)\exp\left[i\upsilon s_x\right]Ai(s_y)\exp\left[i\upsilon s_y\right], \quad (4)$$

where $s_x = x/x_0$ and $s_y = y/x_0$, with the characteristic length in both directions assumed to be equal.

As mentioned in the introduction, an ideal Airy beam exhibits infinite size and thus carries infinite energy, requiring truncation to restrict its power so that it can be physically realized. In the presence of truncation an Airy beam is no longer perfectly diffraction resistant. The specific truncation method matters a lot to its diffraction-free range.

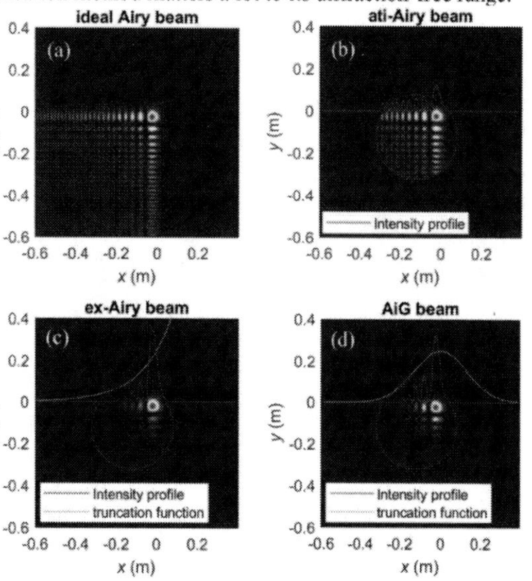

Figure 1. Intensity patterns at the initial plane of (a) ideal Airy beam without truncation, (b) aperture-truncated Airy beam, (c) exponentially decaying Airy beam and (d) Airy-Gauss beam.

In FSOC systems, the finite TX aperture forms a natural boundary, beyond which no beam wave field can exist. An ideal Airy beam truncated by the TX optics is called aperture-truncated ideal Airy (ati-Airy) beam, and its field in the initial plane can be written as

$$U_{\mathrm{ati}}(s_x,s_y) = Ai(s_x)\exp(i\upsilon s_x)Ai(s_y)\exp(i\upsilon s_y)\times \mathrm{circ}(R), \quad (5)$$

where circ () is a circle binary window and R is the radius of the TX aperture.

Considering the rectangular shape of an Airy beam, to make full use of the circular aperture which is most common, we may not put the main lobe of the airy beam right in the center of the aperture. A feasible layout is illustrated in Figure 1 (a).

The most intensively studied finite energy Airy beam so far, however, is the exponentially decaying Airy (ex-Airy) beam, which in the initial plane can be expressed as

$$U_{\mathrm{ex}}(s_x,s_y) = Ai(s_x)\exp\left[(\alpha+i\upsilon)s_x\right]Ai(s_y)\exp\left[(\alpha+i\upsilon)s_y\right], \quad (6)$$

where α is the exponential truncation factor, used to ensure spatial convergence of energy.

Another way to truncate an infinite Airy beam is to apply a Gaussian apodization, which is called the Airy-Gauss (AiG) beam:

$$U_{\mathrm{AiG}}(s_x,s_y) = Ai(s_x)\exp\left[i\upsilon s_x\right]\times Ai(s_y)\exp\left[i\upsilon s_y\right]\exp\left[-r^2/\omega^2\right], \quad (7)$$

where $r = \sqrt{s_x^2+s_y^2}$ and ω is the width of the Gaussian apodization function.

Since propagation distance is the major drawback of an FSOC system based on Airy beams, it is necessary to find out which truncation method in (5-7) enables the largest diffraction free range for a given TX aperture size and a constant Airy characteristic length. There is an analytic formula to calculate the non-diffraction range of an aperture truncated Airy beam [11], but there only the 1D case is considered and it is difficult to generalize to a 2D scenario where the Airy beam is truncated by a *circular* aperture. Also in [11], the authors have demonstrated that the expression for predicting the maximum propagation distance of ex-Airy beams in [12] is incorrect. As for the AiG beam, no prediction for the diffraction free range is available at all.

Here, we try to find out the maximum diffraction-free range of truncated Airy beams with characteristic length x_0 = 25 mm, and the TX pupil radius is set as R = 200 mm. In order to make a fair comparison between the three truncation methods, the truncation parameters of the ex-Airy beam and the AiG beam should be adjusted so that the presence or absence of the specific aperture barely causes extra power loss. This is important because aperture truncation exists anyway for any source beam. Further, the optical power of all the beams in the launching plane is made equal. With the above considerations, the launching fields within the aperture are shown in Figure 1 (b-d), where the intensity distribution of an ideal, untruncated Airy beam is also given as a reference in Figure 1 (a).

For more flexibility, we use the multi-step wave optics simulation (WOS), which is a numerical method based on the FFT version of the Fresnel diffraction integral [13], to

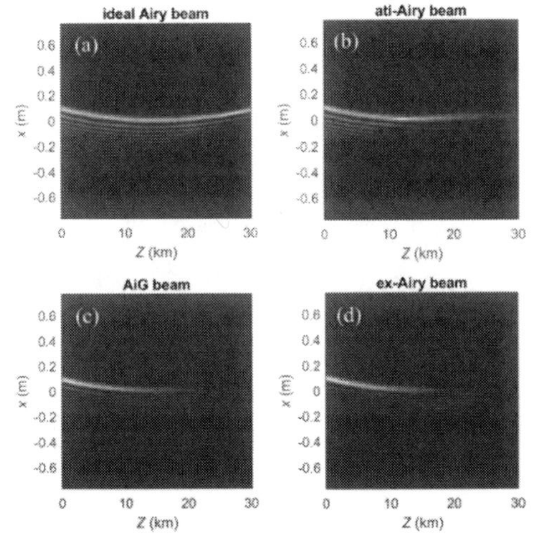

Figure 2. Evolution of intensity profiles in the x-z plane of (a) ideal Airy beam without truncation, (b) aperture-truncated Airy beam, (c) exponentially decaying Airy beam and (d) Airy-Gauss beam.

simulate the propagation process of those truncated Airy beams. Here, the diffraction free range is defined as the distance from the initial plane at which the full width at half maximum (FWHM) of the intensity peak of the main lobe doubles. And the FWHW of the main lobe is measured by its width on the diagonal of the quadrate Airy pattern. The evolution of intensity cross-section of the four Airy beams from Figure 1 is shown in Figure 2, and the corresponding measured FWHMs are given in Figure 3. It is clearly seen that apart from the ideal Airy beam, all the truncated beams are subjected to diffractive spreading beyond a certain distance. In comparison with the ex-Airy beam and the AiG beam, the ati-Airy beam has an obviously longer range of nearly diffraction-free propagation, which is about 27 km in this case, well consistent with the conclusion drawn in [11]. The other two counterparts, however, rapidly collapse beyond 18 km, and are thus excluded from further discussions below.

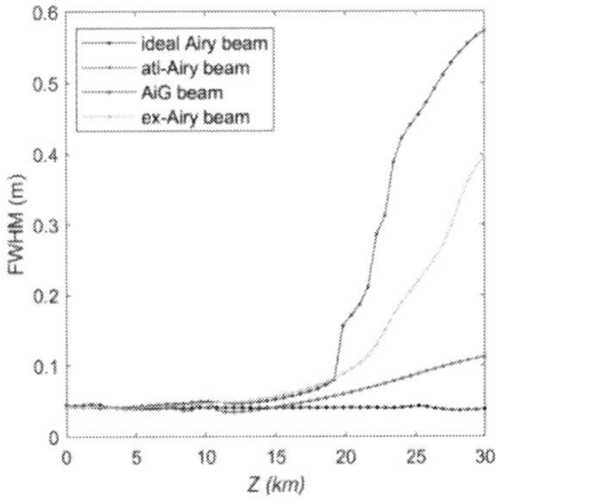

Figure 3. FWHM as a function of propagation distance of Airy beams with different truncations.

III. Effects of Self-Acceleration and Self-Healing on Scintillation in Atmospheric Turbulence

In Section II, we have made it clear that in terms of propagation distance, the ati-Airy beam performs better as an optical carrier than the ex-Airy beam and the AiG beam. And in this section we will find out how well ati-Airy beam's self-acceleration and self-healing act in turbulent atmosphere.

A. Self-acceleration

Without external refractive-index potential along the whole optical path, the acceleration of lateral shift of an Airy beam is only dependent on the Airy characteristic length x_0. If x_0 is constant then in a specified meridian plane, there is only one possible initial velocity/launching angle for the main lobe to connect the transmitter with a certain receiver. But since in the 3D space there are numerous meridian planes, circumvents are possible if we just rotate the 2D launching beam pattern in the launching plane. Under conditions that the transceivers suffer from vibrations, dynamically controlled drop points can be useful in fast tracking.

Now, we are interested in that whether the dynamic control of the ballistic trajectory would affect the turbulence reception performance. To this end, we will examine the scintillation index of an ati-Airy beam in a 20 km horizontal link. The transmitting beam is the ati-Airy beam shown in Figure 1(b), with $x_0 = 25$ mm, and its main lobe adjusted to (0.1 m, 0.1 m) position in the launching plane to fully utilize the aperture with radius $R = 200$ mm. The drop point of the main lobe is controlled by υ through (3), ranging from $x \in (0, 0.5$ m) in the receiver plane, and the free-space trajectories of the main lobe are plotted in Figure 4(a). The atmospheric channel is implemented through 17 evenly distributed phase screens, sampled by 512 points with 3 mm sampling intervals in each side. The scintillation is calculated by the normalized intensity flux variance of a $D = 20$ mm receiver centered on the main lobe intensity peak, and the results are shown in Figure 4(b). The minor fluctuations in the curve are mainly due to simulation inaccuracies, and it is safe to conclude that for truncated Airy beams, the scintillation behavior is not affected by the initial velocity control.

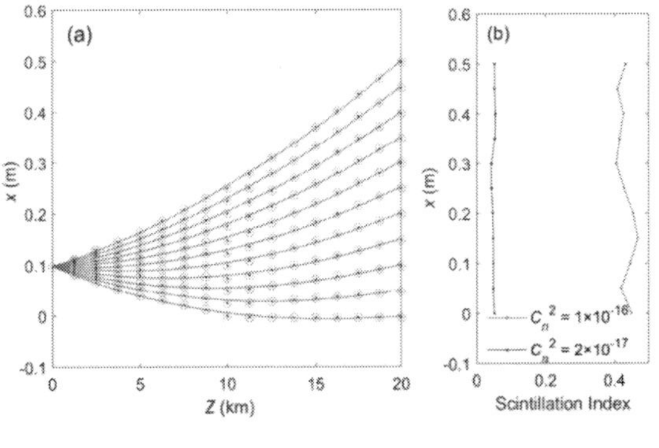

Figure 4. Ati-Airy beams with varied initial velocities. (a) trajectories of the main lobe and (b) scintillation measured at the drop point on the receiver plane.

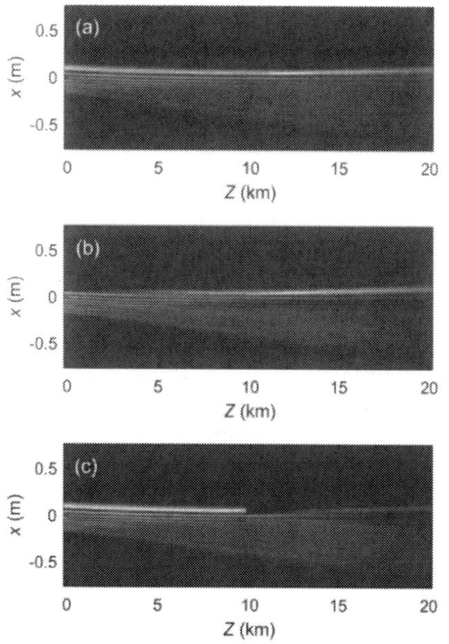

Figure 5. Evolution of intensity profiles of an ati-Airy beam (a) without obstruction, (b) with the main lobe obstructed at the initial plane and (c) with the main lobe obstructed at 10 km.

B. Self-healing

Then we look into the self-healing of ati-Airy beams. Figure 5 illustrates the self-healing process of ati-Airy beams. Figure 5(a) shows an unobstructed ati-Airy beam propagating in free space, while in Figure 5(b) the main lobe is obstructed in the initial plane, and in Figure 5(c) the obstruction in applied in the middle of the path, i.e. 10 km. We see that when the main lobe is obstructed in the initial plane, it is almost fully recovered at the receiver end. By comparison, if the obstruction happens halfway, the recovered main lobe seems much weaker in intensity. One reason for this is because in halfway the side lobes have already dispersed and thus are not able to provide as much power flow. The other reason is that the recovery distance in the case of Figure 5(c) is shorter than that in Figure 5(b). Anyway, in either case the receiver, located at the position where the main lobe is supposed to be, is able to detect the recovered signal so that the link will not fail.

Next, we study the scintillation behavior of the ati-Airy beams with obstructions located at positions as shown in Figure 5(b, c). Nevertheless considering the fact that not all obstacles are totally opaque, we introduce the transparency of the obstacle as a variable τ, and the shape of the obstacle is a rectangle that just covers the main lobe of the Airy beam, see Figure 6 for the effect and shape of the obstacle in the initial plane.

We will begin with a 20 km horizontal link then a 20 km vertical link. The horizontal atmospheric link parameters for simulation are the same as those used in Figure 4(b), only that

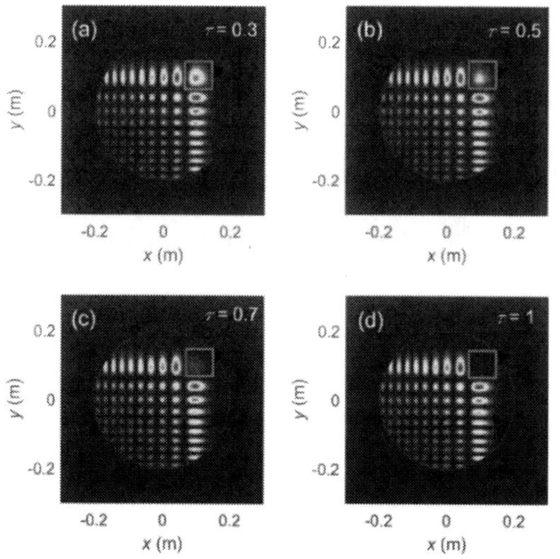

Figure 6. Shape and effect of the obstacle at the launching plane with transparency (a) $\tau = 0.3$, (b) $\tau = 0.5$, (c) $\tau = 0.7$ and (d) $\tau = 1$.

the lateral shift is set to be zero. Three different receiver (RX) aperture sizes are considered, namely, 20 mm, 50 mm, and 100 mm, all centered at the intensity peak of the receiving main lobe. Simulation results are shown in Figure 7.

It seems neither the scintillation behavior nor the mean SNR is greatly affected, indicating that the self-healing process has perfectly counteracted the effects of the obstacle obstructing the main lobe in the launching plane. Similar results can be seen in Figure 8 in which case the main lobe is obstructed at 10 km, i.e. halfway of the path.

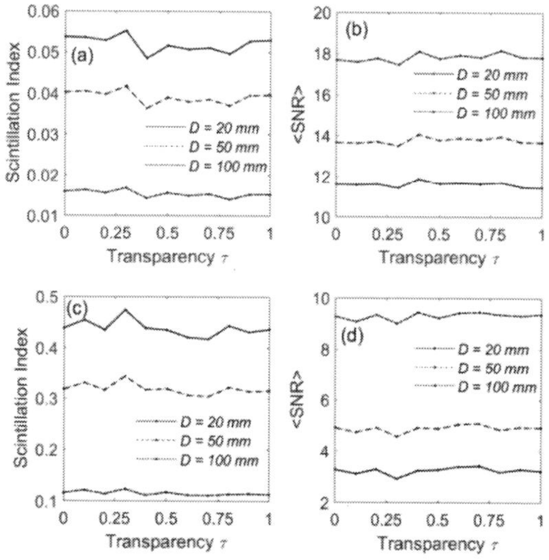

Figure 7. With the main lobe obstructed in the launching plane, aperture-averaged scintillation and mean SNR of ati-Airy beams as a function of the obstacle transparency τ in a 20 km horizontal link with (a, b) $C_n^2 = 2\times10^{-17}$ and (c, d) $C_n^2 = 1\times10^{-17}$.

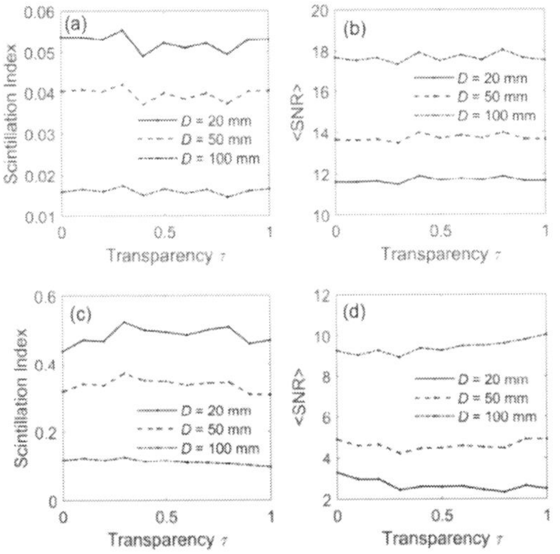

Figure 8. With the main lobe obstructed at 10 km, aperture-averaged scintillation and mean SNR of ati-Airy beams as a function of the obstacle transparency τ in a 20 km horizontal link with (a, b) C_n^2 = 2×10^{-17} and (c, d) C_n^2 = 1×10^{-17}.

Until now we have investigated Airy beams propagating along a horizontal path. Taking advantage of the flexibility of wave optics simulation, we can further examine the performance of Airy beams in turbulent vertical paths. We simulated the case of a 20 km vertical turbulence channel where the H-V$_{5\text{-}7}$ model [14] is applied with the ground-level turbulence set to C_n^2 (0) = 1×10^{-14}. To account for the altitude dependent C_n^2 profile, 17 nonuniformly distributed turbulence phase screens are used for accurate simulation of the channel. The obstacle is put at the halfway of the path, i.e. 10 km, obstructing the main lobe. The results are shown in Figure 9, from which we can see that in the uplink, if the obstacle is translucent then the scintillation will increase a bit but if it is totally opaque, the scintillation even decreases compared with the case of no obstacle (τ = 0), and the resultant mean SNR can increase by 1 dB. While for the downlink, highly opaque obstacles will lead to increase scintillation and decreased SNR.

In general, the self-healing of the ati-Airy beam makes it resistant to obstructions in its propagation path, as a result of which the reception performance is not heavily degraded. This can be explained this way: when the main lobe is blocked, the power of the adjacent side lobes begins to flow to the vacancy. Note that this self-healing flow of optical power comes from a larger area than that occupied by the original main lobe, giving rise to some area averaging in scintillation. This type of averaging also takes place in the propagation of self-focusing beams, such as RPCBs (Radial Partially Coherent Beams) [15], working in essentially the same mechanism with the well-known aperture averaging effect.

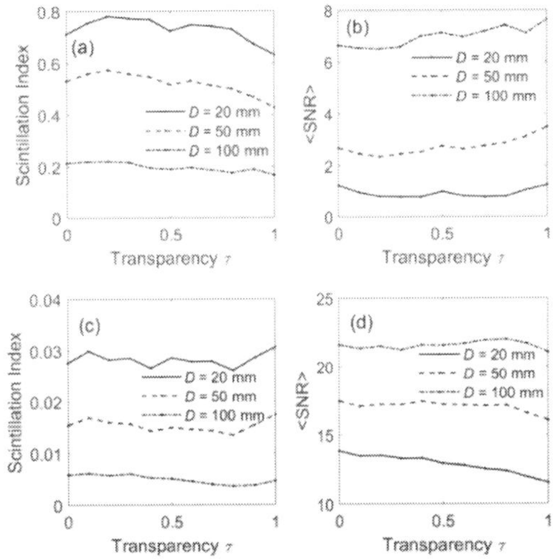

Figure 9. With the main lobe obstructed at 10 km, aperture-averaged scintillation and mean SNR of ati-Airy beams as a function of the obstacle transparency τ in a 20 km (a, b)uplink and (c, d) downlink.

As can be seen from the results in this section, the self-acceleration and self-healing of truncated Airy beams are barely affected by the presence of turbulence. Therefore, these beneficial properties can be exploited in atmospheric free-space laser communication channels.

IV. CONCLUSION AND FUTURE WORK

To investigate the feasibility of applying Airy beams in FSOC systems, we have firstly compared the diffraction-free range of three commonly used truncated Airy beams. With the presence of a hard aperture, an ideal Airy beam with no additional truncation (i.e., the ati-Airy beam) can propagate longer than the exponentially decaying Airy beam and the Airy-Gauss beam, making it more preferable for long-distance FSOC links. Then, the scintillation behavior of ati-Airy beams is studied when the lateral shift acceleration is controlled. It is found that varying the initial velocity will not obviously affect the reception quality. Moreover, the area scintillation and mean SNR of ati-Airy beams in the presence of an obstacle with varying transparency is examined. According to the simulation results, if either the main lobe of the beam is obstructed in the launching plane or in halfway, either in a horizontal or a vertical path, the reception performance maintains quite well. This strongly indicates that, the self-healing feature of an ati-Airy beam makes it resistant to obstacles blocking the propagation path and can thus benefit the link robustness a lot.

In the following research, we will investigate the feasibility of applying truncated symmetrical Airy beams in long distance free-space laser communications.

REFERENCES

[1] L. C. Andrews and R. L. Philips, Laser Beam Propagation Through Random Media, 2nd ed, SPIE, Bellingham, WA, 2005.

[2] C. C. Murat Uysal, Z. Ghassemlooy, A. Boucouvalas, E. Udvary (Eds.), Optical Wireless Communications - An Emerging Technology, Springer International Publishing, Switzerland, 2016.

[3] X. Ji, H. Chen, and G. Ji, "Characteristics of annular beams propagating through atmospheric turbulence along a downlink path and an uplink path," Appl. Phys. B, vol. 122, pp. 221, 2016.

[4] I. Ituen, P. Birch, C. Chatwin, and R. Young, "Propagation of Bessel beam for ground-to-space applications," in Propagation through and Characterization of Distributed Volume Turbulence and Atmospheric Phenomena, Optical Society of America, 2015, PM3C. 4.

[5] G. Gbur, "Partially coherent beam propagation in atmospheric turbulence [Invited]," J. Opt. Soc. Am. A, vol. 31, pp. 2038-2045, 2014.

[6] M. Wang, X. Yuan, J. Li, and X. Zhou, "Radial partially coherent beams for free-space optical communications," in Laser Communication and Propagation through the Atmosphere and Oceans VI, International Society for Optics and Photonics, 2017, pp. 1040813.

[7] G. A. Siviloglou and D. N. Christodoulides, "Accelerating finite energy Airy beams," Opt. Lett., vol. 32, pp. 979-981, 2007.

[8] Y. Hu, G. A. Siviloglou, P. Zhang, N. K. Efremidis, D. N. Christodoulides, and Z. Chen, "Self-accelerating Airy Beams: Generation, Control, and Applications," Springer 170, pp. 1-46, 2012.

[9] G. A. Siviloglou, J. Broky, A. Dogariu, and D. N. Christodoulides, "Ballistic dynamics of Airy beams," Opt. Lett., vol. 33, pp. 207-209, 2008.

[10] J. Broky, G. A. Siviloglou, A. Dogariu, and D. N. Christodoulides, "Self-healing properties of optical Airy beams," Opt. Express, vol. 16, pp. 12880-12891, 2008.

[11] X. Chu, S. Zhao, and Y. Fang, "Maximum nondiffracting propagation distance of aperture-truncated Airy beams," Opt. Commun., vol. 414, pp. 5-9, 2018.

[12] J. Rogel-Salazar, H. Jiménez-Romero, and S. Chávez-Cerda, "Full characterization of Airy beams under physical principles," Phys. Rev. A, vol. 89 (2), pp. 023807, 2014.

[13] D. G. Voelz, Computational Fourier Optics: A MATLAB Tutorial, SPIE Press, Bellingham, Washington, USA, 2011.

[14] R. R. Beland, "Propagation through atmospheric optical turbulence," Atmospheric Propagation of Radiation, vol. 2, pp. 157-232, 1993.

[15] M. Wang, X. Yuan, and D. Ma, "Potentials of radial partially coherent beams in free-space optical communication: a numerical investigation," Appl. Opt., vol. 56, pp. 2851-2857 2017.

Reliable Routing and Spectrum Allocation over Network Coding enabled Elastic Optical Networks

Xin Wang, Yuefeng Ji, Lin Bai
Beijing University of Posts and Telecommunications
Beijing, China
e-mail: xinwang@bupt.edu.cn; jyf@bupt.edu.cn; bailin@bupt.edu.cn

***Abstract*—As each optical link carries a huge amount of hybrid traffic requests, failure of any link may cause huge data loss. In this paper, we mainly discuss the reliable Routing and Spectrum Allocation (RSA) problem over network coding enabled Elastic Optical Network (EON) for hybrid types of traffic requests. As the link fails, a new routing will reconfigure to establish the backup-protection connection for the traffic request. The nodes capable of network coding over network coding enabled EON change the traditional RSA mechanism, which makes the routing processes more complex. To address this problem, we propose a mathematical model to formulate such problem, with the purpose of utilizing the minimal spectrum slots to serve the traffic requests. Besides, a heuristic algorithm called "Reliable Multipath RSA (RM-RSA)" is proposed to solve the model with multipath strategy. For different types of traffic requests, unicast or multicast requests, RM-RSA could provide different routing strategies. The simulation is conducted to verify the efficiency of the proposed RM-RSA, as the changes of average number of multicast receiver, network size, and number of traffic requests. Compared with the benchmark algorithm, the proposed RM-RSA shows more efficient in the spectrum utilization.**

Keywords-reliable RSA; hybrid services;multipath; network coding; EON.

I. INTRODUCTION

The tremendous bandwidth demand triggered by the new emerging high bandwidth services over the optical networks calls for a flexible and reliable network architecture [1]. Each spectrum link carries a large amount of data, and any link failure may cause data loss of 1Gb/s or 1 Tbit/s. How to guarantee the reliable and efficient transmission of service requests becomes urgent to solve, when the network breaks down. Network coding has the natural characteristic of reliable transmission by encoding and encrypting information. Network coding enabled Elastic Optical Network (EON) is the ideal support to accommodate the hybrid requests in the real optical networks [2][3].

We review recent research on the area of reliable and efficient RSA. A new survivable multipath provisioning scheme for unicast services is proposed in EON [4][5]. An Integer Linear Programming (ILP) model and a shared protection virtual optical network mapping algorithm based on different dynamic cost model is also proposed [6]. Single transmission type of service is considered in the research of reliable Routing and Spectrum Allocation (RSA) above. However, various types of services exist in the real optical networks. Survivable multipath routing of two types of traffic demands, anycast and unicast, is studied in EONs [7]. However, none of the research above is to accommodate hybrid services reliable and efficient in network coding enabled EON. The routing becomes more complex as the nodes capable of network coding, and narrow granularity of spectrum resource poses more challenge on spectrum allocation over network coding enable EON.

This paper mainly investigates the reliable RSA problem over network coding enabled EON for hybrid services. The rest of this paper is organized as follows. Section II describes the reliable RSA problem and presents the mathematical model for the problem. Section III proposes an efficient heuristic algorithm to solve the model. Section IV shows the simulation results and makes comparisons of the performances of different algorithms. Section V concludes the paper and presents the future work. The acknowledgement closes the article.

II. RELIABLE RSA PROBLEM

Network coding enabled EON could be a strong support network to provision the hybrid services of multicast and unicast reliable and efficient. This section mainly discusses the reliable RSA problem over network coding enabled EON. A mathematical model is proposed to formulate the problem. Table I shows the parameter definitions of the model. The network coding enabled EON is represented by $G(V,E)$. Given a set of hybrid services, multicast services $R_M^i = \{s_i, D_i, n_i\}, i \in nR_M$ and unicast services $R_U^i = \{s_i, d_i, n_i\}, i \in nR_U$, the object of the mathematical model is to accommodate the hybrid services in a reliable way with the minimal spectrum resources, considering the constraints.

In the routing stage, the nodes capable of network coding change the routing process for multicast service. Compared with routing in EON of establishing one end-to-end path for each source and destination node pair, at least two link-disjoint paths should be established from the source node to each destination node in multicast group [8]. Figure 1 shows the topology of reliable routing for multicast over network coding enabled EON, and multipath transmission strategy is adopted to accommodate multicast services. For each source and destination node pair $(s_i, d_{i,k})$, $w(w=2)$ link-disjoint parallel paths are established as the working path $P_{d_{i,k}}$, representing as the solid lines. Each working path should be provided with a link-disjoint backup protection path, representing as the dotted lines. Note that it has been proved that the best spectrum utilization can be realized only as the number of link-disjoint parallel paths for each source and destination node pair is less than 3 [5]. Thus, w is set as 2 in the proposed model. Besides, it is reasonable to limit the value of w, so as to establish the network coding based multicast tree in the topology. Figure 2 shows the reliable routing topology of unicast over network coding enabled EON. The unicast service is transmitted through multiple paths P_{d_i} and P_{d_i} is provided a link-disjoint backup protection path set $\overline{P_{d_i}}$.

In the spectrum allocation stage, narrow granularity of spectrum resources requires spectrum contiguity constraint and non-overlapping spectrum constraint. Note that we consider the guard band as a default in the spectrum requirement of service request, and guard band constraint is not constrained in the model.

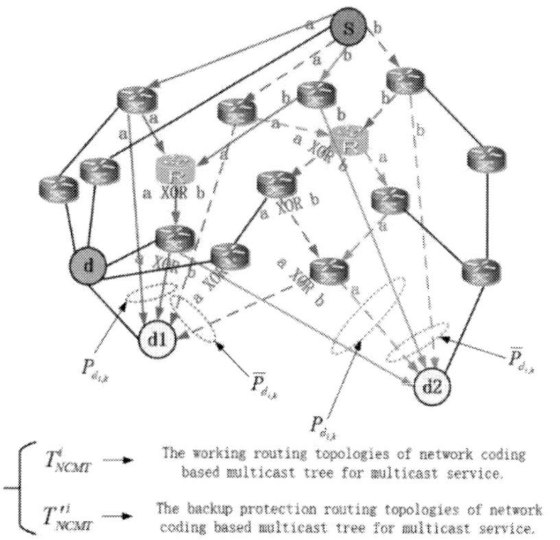

Figure 1. The topology of reliable routing for multicast over network coding enabled EON.

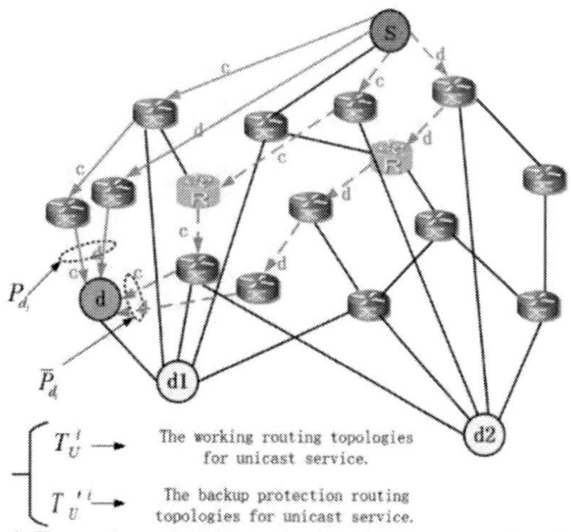

Figure 2. The topology of reliable routing for unicast over network coding enabled EON.

TABLE I. THE PARAMETER DEFINATION OF THE MATHEMATICAL MODEL

Parameter	*Description*
$G(V,E)$	The physical topology of EON.
V	The node set of EON.
E	The edge set of EON.
nR_M	The index set of multicast requests.
nR_U	The index set of unicast requests.
$R_M^i = \{s_i, D_i, n_i\}$	The i-th multicast request set.
$R_U^i = \{s_i, d_i, n_i\}$	The i-th unicast request set.
s_i	The source node of the i-th request.
$d_{i,j}$	The j-th destination node of multicast D_i.
$P_{d_{i,j}}$	The working path set from s_i to $d_{i,j}$.
$\overline{P_{d_{i,j}}}$	The backup protection path set from s_i to $d_{i,j}$.
$p_{d_{i,j}}^k$	The k-th path in path set.

- Minimize:

$$\sum_{i\in nR_M}\sum_{\substack{p\in P_{d_{i,j}}\\ j\in[1,|D_i|]}}\sum_{e\in E}\sum_{f\in[1,F]} O_p^f\cdot\sigma_p^e\cdot\theta_i^p + \sum_{i\in nR_M}\sum_{\substack{p\in \overline{P_{d_{i,j}}}\\ j\in[1,|D_i|]}}\sum_{e\in E}\sum_{f\in[1,F]} O_p^f\cdot\sigma_p^e\cdot\theta_i^p + \sum_{i\in nR_U}\sum_{p\in P_{d_i}}\sum_{e\in E}\sum_{f\in[1,F]} O_p^f\cdot\sigma_p^e\cdot\theta_i^p + \sum_{i\in nR_U}\sum_{p\in \overline{P}_{d_i}}\sum_{e\in E}\sum_{f\in[1,F]} O_p^f\cdot\sigma_p^e\cdot\theta_i^p \quad (1)$$

Equation (1) represents the objective function of the mathematical model, aiming to minimize total spectrum slots of hybrid services occupied on both working paths and backup protection paths. There are some variables needed to be defined. Where O_p^f is the boolean variable, and if the f-th spectrum slot is occupied by path p, O_p^f =1; otherwise, O_p^f =0. σ_p^e is the boolean variable, and if the link e is used by the path p, σ_p^e=1; otherwise, σ_p^e=0. θ_i^p is the boolean variable, and if the path p is used to serve the reqeust R_i, θ_i^p=1; otherwise, θ_i^p=0.

- Constraints:

Equations (2) and (3) ensure the establishment constraints of network coding based multicast tree (NCMT) for multicast request R_M^i. Note that, the number of link-disjoint parallel paths for $(s_i, d_{i,j})$ should be less than 3 so as to achieve the most optimal resource utilization [5]. The number of multiple link-disjoint parallel paths w is set to 2 in this paper, represented as (2).

$$\sum_{p\in P_{d_{i,j}},\overline{P_{d_{i,j}}}} \theta_i^p = w, \forall i \in nR_M \quad (2)$$

$$\sigma_{p_m}^e \cdot \sigma_{p_n}^e \leq 0, \forall p_m, p_n \in P_{d_{i,j}} \cup \overline{P_{d_{i,j}}}, \forall e \in E \quad (3)$$

Equations (4) and (5) ensure the routing constraints for unicast request R_U^i, and the source node and destination node is connected with w link-disjoint paths. Here, w is set to 2.

$$\sum_{p\in P_{d_i},\overline{P_{d_i}}} \theta_i^p = w, \forall i \in nR_U \quad (4)$$

$$\sigma_{p_m}^e \cdot \sigma_{p_n}^e \leq 0, \forall p_m, p_n \in P_{d_i} \cup \overline{P_{d_i}}, \forall e \in E \quad (5)$$

Equations (6) and (7) ensure that the spectrum contiguity constraint [4] on the working paths and backup protection paths for the multicast and unicast requests, respectively.

$$\sum_{f'\in[f+2,F]} O_{p_{d_{i,j}}^k}^{f'} \leq \left(O_{p_{d_{i,j}}^k}^{f} - O_{p_{d_{i,j}}^k}^{f+1} - 1\right)\cdot(-F), \forall p_{d_{i,j}}^k \in P_{d_{i,j}}, \overline{P_{d_{i,j}}}, j \in [1,|D_i|], \forall f \quad (6)$$

$$\sum_{f'\in[f+2,F]} O_{p_{d_i}^k}^{f'} \leq \left(O_{p_{d_i}^k}^{f} - O_{p_{d_i}^k}^{f+1} - 1\right)\cdot(-F), \forall p_{d_i}^k \in P_{d_i}, \overline{P_{d_i}}, \forall f \quad (7)$$

Equations (8) and (9) ensure the non-overlapping spectrum constraint [4] on the working paths and backup protection paths for the multicast requests and the unicast requests, respectively. Where λ_p^f is the boolean parameter, and if the f-th spectrum slot is occupied on every link of path p, λ_p^f=1; otherwise, λ_p^f=0.

$$\lambda_{p_{d_{i,j}}^k}^f \cdot O_{p_{d_{i,j}}^k}^f \leq 0, \forall f, \forall p_{d_{i,j}}^k \in P_{d_{i,j}}, \overline{P_{d_{i,j}}}, j \in [1,|D_i|] \quad (8)$$

$$\lambda_{p_{d_i}^k}^f \cdot O_{p_{d_i}^k}^f \leq 0, \forall f, \forall p_{d_i}^k \in P_{d_i}, \overline{P_{d_i}} \quad (9)$$

As we research some related references, most mathematic models of RSA problem are formulated over the EON [4][7]. A new survivable multipath provisioning scheme is proposed and an ILP model is developed for the scheme in [7]. However, the proposed model just considers one type of service, unicast. Although an ILP model of survivable multipath routing is proposed for hybrid services of anycast and unicast traffic, multicast services are not served over the EON. Due to that the routing process is quite different in network coding enabled EON, as demonstrated in Equations (2)-(5), these models can not be applied into our problems. Although the model of hybrid services efficient provision problem in [2] is proposed in network coding enabled EON, reliable connection is not guaranteed as link failure occurs. In our proposed model, reliable of RSA over network coding enabled EON is formulated for hybrid services of unicast and multicast services.

III. PROPOSED ALGORITHMS

A. *Reliable Multipath RSA*

Due to that the mathematical model is not efficient to solve the reliable RSA in large networks, an efficient heuristic algorithm, called "Reliable Multipath Routing and Spectrum Allocation (RM-RSA)", is proposed. The detailed procedures of RM-RSA are shown below.

Step 1: To initialize the NCMT set T_{NCMT}^i, the backup protected NCMT set $T_{NCMT}^{\prime i}$ for multicast service R_M^i, unicast routing set T_U^i, backup protected unicast routing set $T_U^{\prime i}$ for unicast service R_U^i, layered graph set $U^p(V^p, E^p)$, and total occupied spectrum slots $Nfstotal$.

Step 2: According to transmission types of unicast or multicast, clarifying the hybrid services into $R_U^i, i \in nR_U$ and $R_M^i, i \in nR_M$;

Step 3: Adopting multicast priority strategy, to process the service request one by one;

Step 4: Subject to the spectrum contiguity constraint of (6), and non-overlapping spectrum constraint of (8), adopting the layered graph approach [9] to establish $U^p(V^p, E^p)$, which meets with the spectrum requirement of R_M^i;

Step 5: Based on $U^p(V^p, E^p)$ obtained in Step 4, utilizing the Warshall-Floyd algorithm [10] to establish $w(w{=}2)$ link-disjoint parallel paths for each $(s_i, d_{i,j})$ of R_M^i as the working paths, subject to the constraints of (2) and (3);

Step 6: To mark the usage status of occupied spectrum slots in B0 as used, and set the weight of occupied link as *Inf*;

Step 7: To establish the backup protection structure of each working path for each $(s_i, d_{i,j})$ of R_M^i as Step 5, and set the spectrum usage status in B0 as reserved;

Step 8: To calculate total occupied number of spectrum slots saved as $Nfstotal$, and initialize the weight matrix of $G(V,E)$;

Step 9: Repeating from Step 4 to Step 8, establishing the reliable NCMT until all the destination nodes $d_{i,j}$, $j \in [1,|D_i|]$ are connected to s_i;

Step 10: To save the established NCMT as T_{NCMT}^i and backup protected NCMT as $T_{NCMT}^{\prime i}$ and calculate total spectrum utilization saved as $Nfstotal$;

Step 11: Repeat from Step 4 to Step 10, until all $R_M^i, i \in nR_M$ are processed;

Step 12: Subject to the spectrum contiguity constraint of (7), and non-overlapping spectrum constraint of (9), adopting the layered graph approach [9] to establish $U^p(V^p, E^p)$, which meets with the spectrum requirement of R_U^i;

Step 13: Based on the obtained $U^p(V^p, E^p)$ in Step 12, utilizing the Warshall-Floyd algorithm [10] to establish $w(w=2)$ link-disjoint parallel paths for (s_i, d_i) of R_U^i as the working paths, subject to the constraints of (4) and (5);

Step 14: To mark the usage status of occupied spectrum slots in B0 as used, and set the weight of occupied link as Inf;

Step 15: To establish the backup protection path for the working path of (s_i, d_i) according to the method in Step 13;

Step 16: To calculate total occupied number of spectrum slots saved as $Nfstotal$, and initialize the weight matrix of $G(V, E)$;

Step 17: To obtain T_U^i and backup protected structure $T_U^{\prime i}$ and calculate total spectrum utilization saved as $Nfstotal$;

Step 18: Repeat from Step 12 to Step 17, until all $R_U^i, i \in nR_U$ are processed;

IV. SIMMULATION RESULTS

In this section, we conducted simulations to evaluate the performance of the proposed algorithms. Two commonly used algorithms select as the benchmark algorithms, which are K shortest paths-first fit algorithm (KSP-FF) and K shortest paths-random fit (KSP-RF). Note that, we considered the reliable routing path for each working path, and the benchmark algorithms become Reliable-KSP-FF (RKSP-FF) and Reliable-KSP-RF (RKSP-RF).

We use the tools of MATLAB to realize the simulation. The spectrum resources over the network coding enabled EON are deployed on the C band, and the total capacity of spectrum link is 4.475THz. The spectrum granularity is assumed as 12.5GHz, and the corresponding number of spectrum slots on each spectrum link provides 358 at most. The index number of source and destination nodes are random selected in the range of $[1, N]$. The number of parallel link-disjoint paths are set as 2. The total occupied number of spectrum slots $Nfstotal$ consider as only parameter to evaluate the spectrum utilization of the proposed algorithm. The performance of the proposed algorithm is evaluated in the test networks of random networks. The impacts of different parameters of average receiver number, network size, and number of hybrid requests are evaluated on the performance of the proposed algorithm in the following. Besides, as serving multicast requests, the performances of the proposed algorithm considering network coding or not are compared.

A. *Impact of average receiver number*

Figures 3 and 4 show that RM-RSA consumes the least number of spectrum slots to accommodate the hybrid services of same scale and different scales among other benchmark algorithms, respectively. As the average receiver number (nD) increases, RM-RSA considering network coding outperforms RM-RSA in the performance of spectrum utilization, and the average spectrum utilization is improved by about 9.8%. This is reasonable because under the layered graph approach, spectrum allocation considers firstly, and routing is established based on a set of layered graph meeting with the spectrum requirement. Besides, network coding could also improve the spectrum utilization of link.

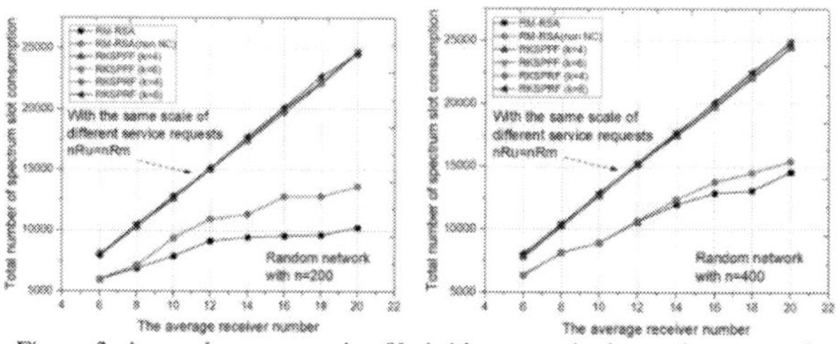

Figure 3. As serving same scale of hybrid requests in the random networks of (a) n=200; (b) n=400, the impact of nD on the spectrum utilization of RM-RSA.

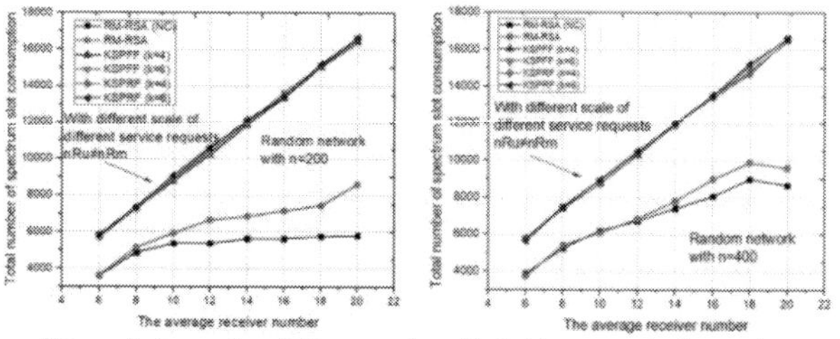

Figure 4. As serving different scales of hybrid requests in the random networks of (a) n=200; (b) n=400, the impact of nD on the spectrum utilization of RM-RSA.

B. *Impact of network size*

As the random network size n increases, the proposed algorithm RM-RSA shows the best spectrum utilization performance among other benchmark algorithms in both random networks of small size and large size shown in Figures 5 and 6. As serving same scale of hybrid services, RM-RSA reduces the average spectrum resources consumption by about 29.64% among other benchmark algorithms. Moreover, RM-RSA shows more efficient, especially in large random networks than in small random networks.

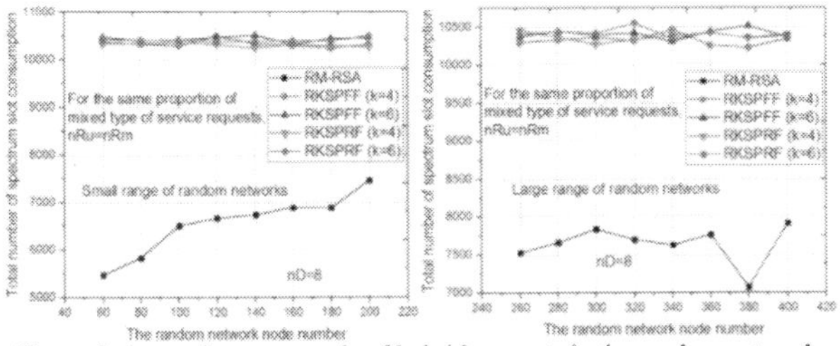

Figure 5. As serving same scale of hybrid requests in the random networks of (a) small size; (b) large size, the impact of n on the spectrum utilization of RM-RSA.

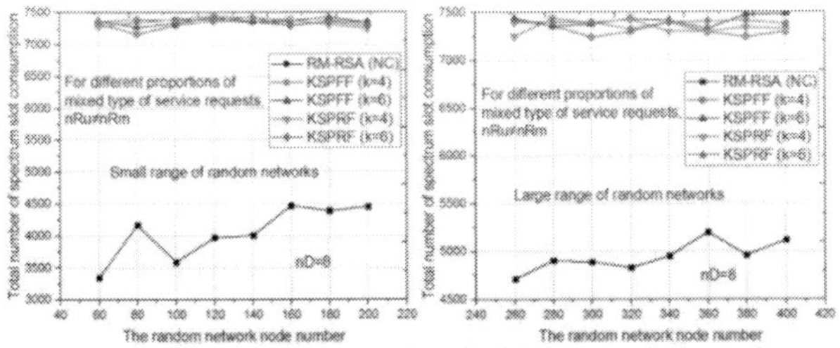

Figure 6. As serving different scales of hybrid requests in the random networks of (a) small size; (b) large size, the impact of n on the spectrum utilization of RM-RSA.

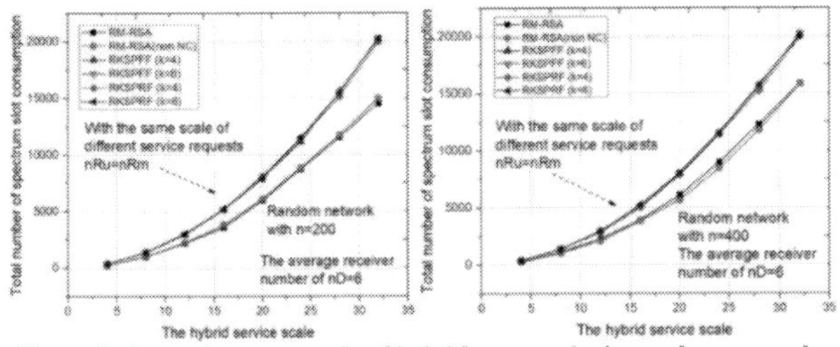

Figure 7. As serving same scale of hybrid requests in the random networks of (a) n=200; (b) n=400, the impact of nR on the spectrum utilization of RM-RSA.

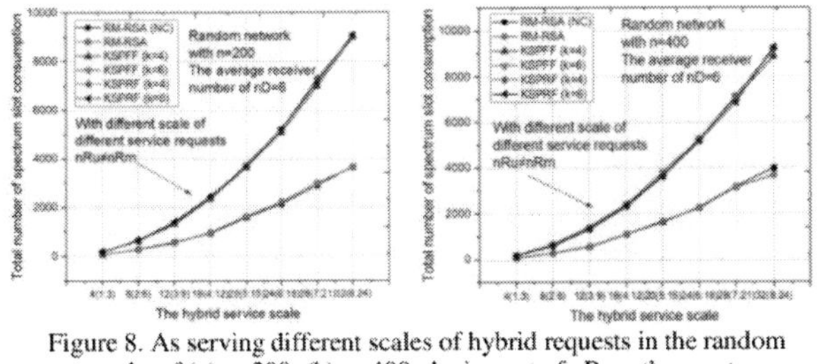

Figure 8. As serving different scales of hybrid requests in the random networks of (a) n=200; (b) n=400, the impact of nR on the spectrum utilization of RM-RSA.

C. *Impact of number of hybrid requests*

As the number of hybrid requests increases, RM shows the best spectrum utilization among other algorithms in both random networks in Figures 7 and 8. Note that, the simulation results above are obtained under the specific simulation conditions, as serving the limited size of hybrid requests.

V. Conclusion and Future Work

In this paper, we mainly investigate reliable multipath RSA for hybrid services of unicast and multicast over network coding enabled EON. We formulate a mathematical model and propose an efficient heuristic algorithm "Reliable Multipath RSA (RM-RSA)". Simulation results show that the proposed algorithm RM-RSA shows the best spectrum utilization among other benchmark algorithms in random networks, as different parameters of average receiver number, network size, and number of hybrid requests change. A large number of simulations need to evaluate the impact of large-scale request services on the performance of the algorithm in the future work. Due to that free space optical (FSO) communication networks have high requirement for reliable transmission, the application of the proposed scheme into the FSO can be a good choice.

Acknowledgment

This work was supported by the National Natural Science Foundation of China (Grant No. 61771073).

References

[1] N. Chowdhury and R. Boutaba, "A survey of network virtualization," Comput. Netw., vol. 54, no. 5, pp. 862–876, Apr. 2010.

[2] X. Wang, R. T. Gu, and Y. F. Ji, "RSA for the hybrid unicast and network coding based multicast services over the flexible optical networks," Proc. of the 25th Wireless and Optical Communication Conference 2016, pp. 1–4, Chengdu, China, May.21-23, 2016.

[3] Y. F. Ji, J. W. Zhang, Y. L. Zhao, X. S. Yu, J. Zhang, and X. Chen, "Prospects and research issues in multi-dimensional all optical networks," SCIENCE CHINA Information Sciences,vol.59,no.10,pp.1–14, 2016.

[4] L. Ruan and Y. Zheng, "Dynamic survivable multipath routing and spectrum allocation in OFDM-based flexible optical networks," Journal of Optical Communications and Networking, vol.6,no.1,pp.77-85, 2014.

[5] L. Ruan and N. Xiao, "Survivable multipath routing and spectrum allocation in OFDM-based flexible optical networks", Journal of Optical Communications and Networking, vol. 5, no. 3, pp. 172-182, 2013.

[6] Y. Y. Wang, X. Li, B. L. Guo, T. Gao, W. Z. Li, and S. G. Huang, "Survivable virtual optical network mapping in elastic optical networks with shared backup path protection," Proc. of the 25th Wireless and Optical Communication Conference 2016, pp. 1–4, Chengdu, China, May.21-23, 2016.

[7] R. Goścień, K. Walkowiak, and M. Tornatore. "Survivable multipath routing of anycast and unicast traffic in elastic optical networks." Journal of Optical Communications and Networking, vol.8,no.6, pp. 343-355,2016.

[8] X. Wang, R. T. Gu, Y. F. Ji, and M. Kavehrad, "Hybrid services efficient provisioning over the network coding enabled elastic optical networks," Optical Engineering, vol.56, no.3, Mar, 2017, pp. 036101(1-14).

[9] X. H. Liu, L. Gong, and Z. Q. Zhu, "Design integrated RSA for multicast in elastic optical networks with a layered approach," Proceedings of 2013 IEEE Global Communications Conference (GLOBECOM), pp.2346-2351,2013.

[10] J. F. Myoupo and A. C. Fabret, "A modular systolic linearization of the Warshall-Floyd algorithm," IEEE Transactions on Parallel and Distributed Systems,vol.7.no.5, pp.449-455,1996.

Free Space Optical Communication Networks

Peng Deng*[1], Tim Kane[1], Xiuhua Yuan[2], Minghao Wang[1,2], Wentao Xia[1]
[1]Department of Electrical Engineering, The Pennsylvania State University, University Park, USA
[2]School of Optical and Electronic Information, Huazhong University of Science and Technology, Wuhan, China
e-mail: *pxd18@psu.edu

Abstract—**Effective integration of high speed, power efficient, compact and reliable free space optical communication terminals with small satellite platforms can pave the way to satellite constellations telecommunication networks. This paper will discuss the current architecture limitations that prevent the integration, and the critical technologies to be addressed for the realization of these systems, including compact wide field of view optical antenna, Micro-Electromechanical System (MEMS) gimbal-less fast acquisition and tracking mechanism, spatial diversity and beam shaping techniques for atmospheric channel. This development would explore the landscape of miniaturized high capacity rugged free space optical communication systems for future space networks.**

Keywords-free space optical communication; small satellties, CubeSats; optical antenna; MEMS; beam steering; acquisition, tracking and pointing; spatial diversity; beam shaping.

I. INTRODUCTION

The rapid growth of satellite constellations, enabled by the increasing availability of small-sized and low-cost satellites such as Low Earth Orbit (LEO) Cubesats, can provide global real-time remote sensing and communication network [1]. Data-intensive satellite sensors produce a large amount of information to be transmitted to the ground in a short time, which requires high-capacity communications. Information security is also becoming an urgent issue in satellite constellations, because the amount of critical and valuable data to be communicated is increasing. Tiny satellites such as nanosats and small microsats may lack the power supply or mass for large conventional radio transponders. Small satellite applications are constrained by their ability to provide high bandwidth secured data communications for highly capable payloads [2].

Free Space Optical Communications (FSO) have evolved as a promising alternative for high-capacity data links from space, due to the high gain of a narrow beam, ultralow inter-channel interference and featuring smaller and lighter terminals [3]. Moreover, FSO communications integrated with a quantum receiver, provide a requisite platform for intrinsically hack-proof secure communication, known as Quantum Key Distribution (QKD) [4]. Free Space Optical Communications can enable high speed multi-access space networks [5] to spread across previously inaccessible platforms including satellite-to-satellite crosslinks, up-and-down links between space platforms and aircraft, ships, among mobile and stationary terminals, and other ground platforms. Figure 1 shows Free Space Optical Communication Network [6].

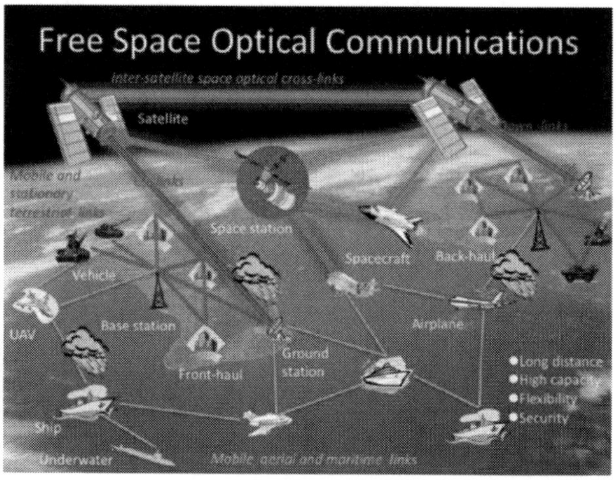

Figure 1. Free Space Optical Communication Network.

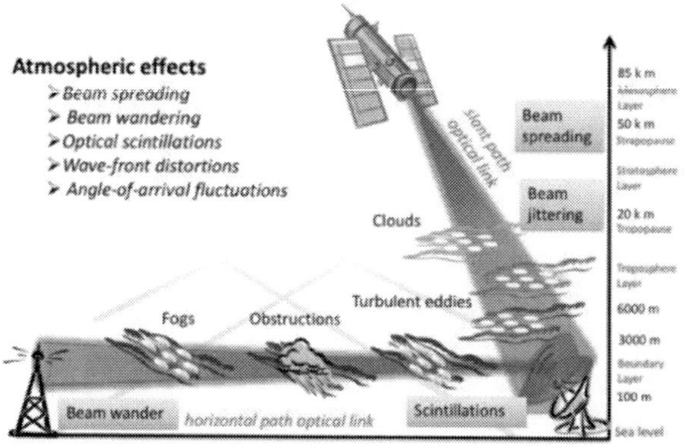

Figure 2. Atmospheric effects on Free Space Optical Communications.

However, traditional FSO system using large size telescope with narrow Field of View (FOV), and power consuming gimbal based Acquisition, Tracking and Pointing (ATP) mechanism [7], cannot meet the requirements of cost and Size, Weight, and Power (SWaP) in nanosatellite applications. Furthermore, atmospheric turbulence induced effects such as scintillations, beam wander and beam jittering can degrade free space optical communication link performance [8][9]. Figure 2 shows atmospheric effects on Free Space Optical Communications [6].

The effective integration of high speed, power efficient, compact and reliable free space optical communication terminals with small satellite platforms can pave the way to satellite constellations telecommunication networks. This paper will discuss the current architecture limitations that prevent the integration of laser communication terminals onboard small space platforms, and explore the critical technologies to be addressed for the realization of these systems. We will discuss compact wide field of view optical antenna [10] in Section II, and MEMS gimbal-less retroreflective acquisition and tracking mechanism [11] in Section III. Section IV presents spatial diversity and beam shaping techniques [12] for atmospheric channel. Fundamental limits of performance and design methodologies for promising future research and development will be given. These technologies would help explore the landscape of miniaturized high capacity rugged free space optical communication systems for the development of future satellites networks.

II. WIDE FIELD OF VIEW OPTICAL ANTENNA

In traditional FSO communication systems, several types of telescopes have been utilized at the front-end of optical antenna, such as Galileo telescopes and Cassegrain telescopes. Schmidt-Cassegrain telescopes are the most popular types of telescopes that eliminate the spherical aberration and compensate for high-order aberration such as coma and astigmatism. It has a slight light loss due to the secondary mirror obstruction compared to refractors. A compact optical antenna using off-axis free form surface triple mirrors [13] was developed to eliminate the obstruction for a high speed laser communication system.

However, the narrow FOV of optical antenna is the inherent characteristic that makes it difficult to maintain the Line-Of-Sight (LOS) links for FSO communications. A compact bidirectional adaptive optical receiver with a wide-angle tracking and adaptive compensation can improve the concentration of laser beam energy and reduce signal fading. Design of a compact adaptive optical receiver utilized fiber-tip positioning systems and piezoelectric bimorph adaptive mirrors for beam steering and low-order wavefront control [14]. In order to observe a large FOV and acquire a wide spatial acquisition range, a correction procedure can be developed [7] from a wide FOV lens imaging model to a pinhole imaging model and used a fisheye lens in a coarse pointing system [15]. In addition, multiple-aperture imaging systems, beam divergence changing mechanisms and an optical phased array receiver can be investigated to improve the optical antenna [16] field-of-view for FSO communications.

Figure 3 shows wide field of view optical antennas for free space optical communications [17]. A wide FOV antenna can be achieved by using a single-element fisheye lens group to collect the wide field beam, and a steering mirror to couple the beam into a multi-mode fiber [10]. The inherent motion environment demands on the accuracy of a tracking mechanical coupling system. Another method to improve the FOV and eliminate the tracking mechanism is to use angle diversity receiver, where an array of narrow-FOV non-imaging receiver elements are oriented along different directions to cover a wide FOV. Diversity schemes such as select-best, equal gain combining [18], and maximal-ratio combining [19] can be employed to improve the received Signal-to-Noise Ratio (SNR) performance. However, implementation of angle diversity using a separate optical receiver for each directional element is excessively bulky and costly. A fly-eye imaging diversity receiver [20], which consists of a single imaging optical concentrator that forms an image of the received light from different directions on a separating element of detector array. Imaging receivers and multi-beam transmitters [21] can reduce the required transmitter power. Moreover, when the multiple beams from multiple receiver elements are spatially separated and independently combined, the effects of intensity fluctuation average out to reduce the probability of signal fades.

An off-axis catadioptric fisheye wide FOV optical antenna for FSO communications [10] is shown in Figure 4. The optical antenna consists of a fisheye lens group and a catadioptric telescope with off-axis aspheric surface mirrors. The optical device elements are numerically analyzed and optimally designed so that the receiver structure can make a positive contribution in enlarging the FOV and reducing optical aberrations. In order to correct the incident beam translation and compensate the beam wandering or beam jittering effects, a double-level tracking mechanism is incorporated into the optical antenna. The purpose of this optical antenna is to provide a 60-degree wide field of view to expand the tracking range, and mitigate optical aberrations to improve the tracking accuracy for FSO systems [10].

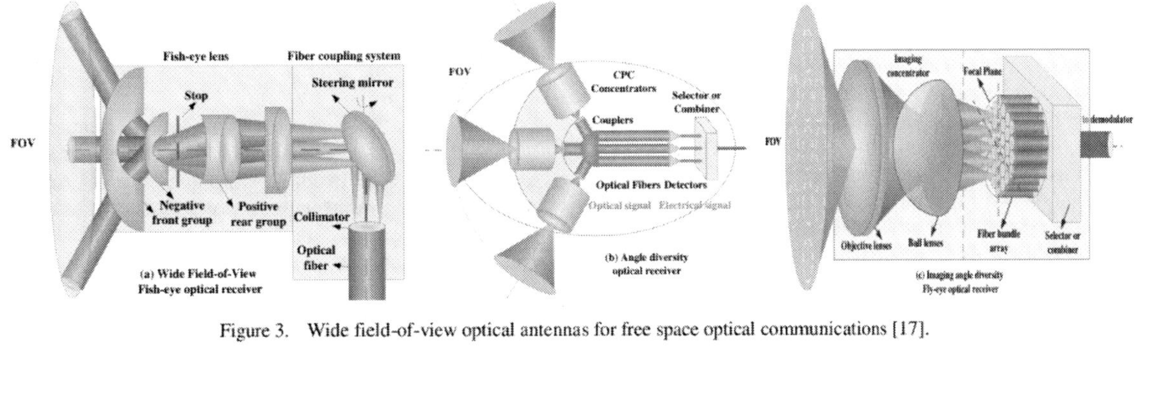

Figure 3. Wide field-of-view optical antennas for free space optical communications [17].

III. MEMS GIMBALESS ACQUISITION AND TRACKING

Free space optical communication system usually employs an ATP mechanism to point the receiver's narrow field of view at the small divergence beam [7]. ATP system, consisting of a coarse pointing system, a fine pointing system, and a point-ahead system, has already been incorporated into a satellite terminal to compensate for the LOS error and dynamic disturbance [5]. However, traditional ATP systems have large size, high power consumption and high cost. The complex coarse pointing system increases weight and volume of optical antenna and reduces tracking efficiency. This approach is impractical for small satellite applications requiring small size and low weight. Thus, a compact and power efficient gimbal-less acquisition and tracking mechanism should be developed for small satellite platforms.

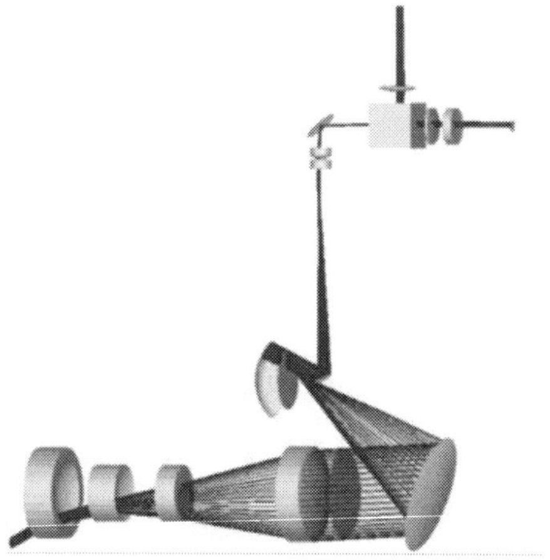

Figure 4. Off-axis Catadioptric Wide field-of-view optical antenna

For acquisition and tracking mechanism, beam steering can be accomplished by changing the refractive index of the medium through which the beam is transmitted or by the use of mirrors, prisms, lenses, or rotating diffraction gratings [22]. Optical beam steering approaches include mechanical mirror-based gimbals, GM mechanisms that rotate mirrors, Risley prisms, phased-array diffraction optics, and Micro-Electromechanical System (MEMS) using micro-mirrors [23]. Conventional mechanical Fast Steering Mirrors (FSMs) using voice coils have large mirrors (25mm) with large moving masses and high power (10W) consumption, which make them more susceptible to shock and vibration. Piezoelectric mirrors offer sufficient steering range (50 mrad) and resolution (5 urad), but piezoelectric actuation exhibits strong nonlinearity and hysteresis which require the use of a complex and large size controller. Galvanometer Mechanisms (GM) have the small size, low cost and large steerable angle (~40 degree), but at the cost of low steering speed (~60 Hz) and high-power consumption (~W). MEMS-based beam steering are much speedier (~kHz), more precise and reliable, and requires a lower power (~mW). Though MEMS beam steering are restricted to small mirrors (1-8mm) or a smaller coverage cone (~10 degree) [23]. We can use a beam expander to adjust beam size or a wide angle field fisheye lens to magnify the optical scan-angles [10]. As a promising beam steering approach within the SWaP constraints, gimbal-less MEMS-based beam acquisition and tracking mechanism can be developed for flexible FSO satellite network.

Figure 5. Experiment Setup of 2D MEMS beam steering mechanism to switch Free Space Optical Communication links [11]

For reconfigurable FSO links, we need a beam acquisition and tracking mechanism that can automatically search for the target receiver position quickly, point the beam from transmitter to receiver aperture, and track the center of aperture precisely. Traditional acquisition and tracking using mechanical mirror-based gimbals and electrical feedback signals from the receiver would be lengthy. In addition, a single FSO device assembly, including beam steering and alignment machinery, should have high efficiency in terms of size, cost and power consumption [22]. For this reason, we proposed a separate visible laser beacon based tracking system that integrated a 2D gimbal-less MEMS mirror with retroreflective optics [22]. This system uses a high sensitive photodetector at the transmitter to detect the reflective light from retroreflective tags placed around the receiver apertures. This concept [22] was recently demonstrated for a MEMS beam steerable FSO link working at a data rate above 10 Gb/s [24] with a field-of-view (FOV) of ±4.2°. Experiment setup of gimbal-less MEMS beam acquisition and tracking mechanism to switch Free Space Optical Communication links for a flexible wireless data center inter-rack network is shown in Figure 5.

With the initial target location, MEMS beam acquisition mechanism auto-search for the target around the initial location within the beam steering angle range. According to various distance and angular directions of retroreflective receiver aperture, we develop intelligent MEMS scanning patterns for beam acquisition and tracking mechanism [11]. The adaptive beam acquisition patterns and tracking algorithm parameters are optimized in term of the acquisition time, searching efficiency and tracking accuracy.

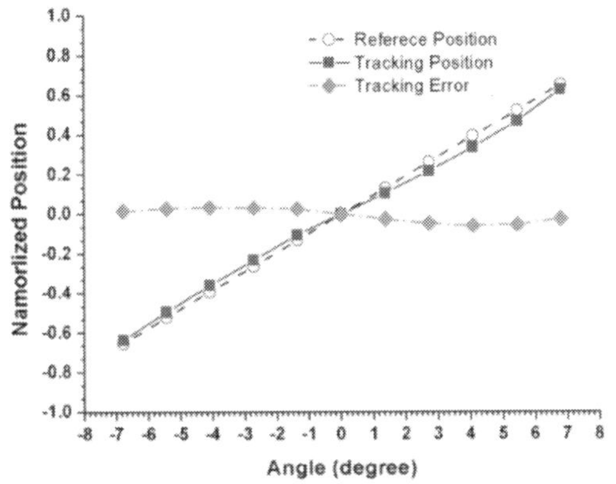

Figure 6. Beam acquisition and tracking position accuracy of MEMS switchable FSO system as function of scan angle [22].

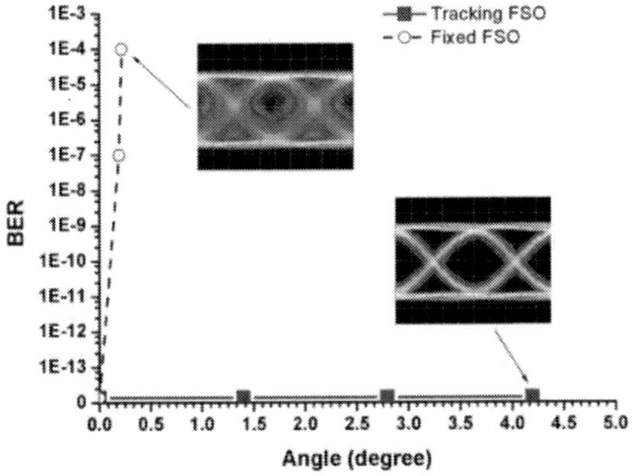

Figure 7. BER performance as function of tracking angle for the fixed FSO link and the MEMS tracking FSO at 10Gbps [22].

An adaptive scan pattern is performed for intelligent acquisition and tracking system, to search and identify the retro-reflective receiver aperture at arbitrary positions within the scanning range of the MEMS mirror [22]. As hitting a retro-reflector, a strong signal peak is detected and associated to the MEMS mirror position. The estimated positions of the retro-reflectors are used as initial values of the reference positions for the line of sight alignment of tracking system [11]. For this experiment, we enhance the distance from MEMS transmitter to retroreflective receiver by using adaptive threshold detection and intelligent acquisition tracking patterns, and improving optical power coupling efficiency. When the retroreflective receiver aperture is moved along a linear transition direction at various distance, beam acquisition and tracking positions and response times are recorded in order to characterize MEMS tracking system with respect to acquisition accuracy and tracking latency. Figure 6 and Figure 7 show tracking position accuracy and BER performance of MEMS beam switchable FSO at 10 Gbps, respectively.

Design and implement of an agile reconfigurable bidirectional 10Gbps FSO system incorporated with intelligent beam acquisition and tracking mechanism is built based on gimbal-less two-axis MEMS micro-mirror and retro-reflective optics [22]. The gimbal-less MEMS reconfigurable FSO links with intelligent adaptive acquisition and tracking are evaluated over various distances and directions to improve the acquisition latency and tracking accuracy [11]. Optical power loss and bit error rate of reconfigurable 10Gbps FSO links with mobility are evaluated over a short distance. The long distance FSO system can be achieved by using high power lasers such a s MOPA and EDFA, high sensitive photodetectors of APD single photon detectors and high gain wide field-of-view optical antennas. Further research work will focus on long distance bidirectional reconfigurable free space optical communication links using advanced MEMS adaptive acquisition and tracking to address environment effects such as LOS blockages, vibrations and intensity fluctuations.

IV. SPATIAL DIVERSITY AND BEAM SHAPING

Intensity fluctuations or scintillation of laser beam propagation through atmospheric turbulence will degrade the BER performance and channel capacity of free space optical communications between satellites and ground platforms. In order to improve communication performance, scintillations can be mitigated by means of spatial diversity using multiple transmitted beams and multiple receivers [25], reducing spatial coherence of the transmitted beam [26] and advance beam shaping techniques.

A. *Spatial diversity and Spatial coherence*

Spatial diversity using multiple transmitted beams and multiple receivers can also be employed to reduce scintillation and ultimately improve FSO channel capacity. It has been shown that the scintillation of a beam array can be reduced by carefully adjusting the spatial separation of beamlets [25][27]. However, scintillation of a beam array will increase significantly if the spatial separation of beamlets is smaller than the correlated length. In addition, the received energy from the beam arrays is low unless the constituent beamlets are inclined to overlap at the receiver aperture, which is difficult to achieve over long propagation distances. The use of multiple transmitters and receivers has also been suggested for use in Multiple-Input–Multiple-Output (MIMO) configurations [27].

Partially coherent beams with reduced spatial coherence show lower scintillation at the cost of larger divergence angle and lower average received power. Partially coherent beams have a lower scintillation than fully coherent beams. However, a partially coherent beam has a larger beam spreading and forms a large spot in the receiver aperture, which leads to a loss of the transmitted energy being received by the detector. By optimizing the spatial coherence length, the improvement in scintillation reduction can overcome the penalty of power reduction and significant signal-to-noise ratio gains can be obtained in weak atmospheric turbulence [28].

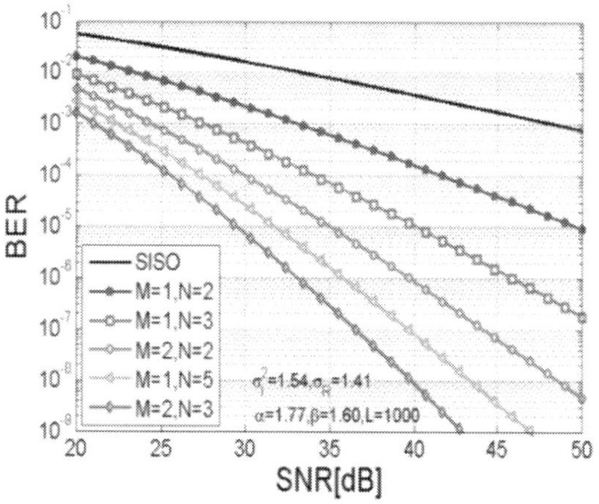

Figure 8. BER performance of MIMO free space optical communications in atmospheric turbulence [17].

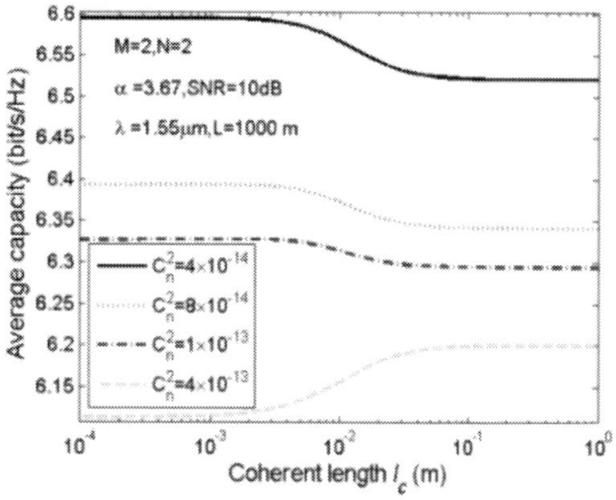

Figure 9. Channel capacity of MIMO free space optical communications using multiple partially coherent beams [12].

However, MIMO FSO links studied so far employ the fully coherent beams whose spatial correlated length is large and the induced scintillation is strong. The mutually independent MIMO branches require a relatively larger separation between the transmitter and receiver apertures than the correlated length. Partially coherent beams with shorter spatial coherence lengths can help reduce the scintillation and correlation length in strong turbulence, and eventually reduce the required separation space between independent apertures [12]. Thus, we proposed to develop MIMO FSO links whose spatially separated beamlets can have lower spatial coherence lengths that lead to reduced scintillation and smaller separation between transmitter and receiver apertures [12]. BER performance of MIMO FSO links in non-Kolmogorov moderate to strong turbulence was evaluated in Figure 8, and the effect of spatial coherence length of partially coherent beams on capacity of MIMO FSO links [12] was analyzed in Figure 9. The approach that incorporates partially coherent beams into the MIMO FSO links could reduce scintillations and correlation length and to improve average channel capacity in weak-media atmospheric turbulence. In strong turbulence, large scale size of turbulent eddy will become shorter than the spatial coherence length, which lead to the increasing large-scale log-irradiance variance and the decreasing average capacity.

B. Beam shaping and Airy beams

Beam shaping is the process of redistributing the irradiance and phase of a beam of optical radiation. The beam shapes is defined by the irradiance distribution and the phase of the shaped beam is a major factor in determining the unique propagation properties of the specific beam profiles, such as Bessel beams, Laguerre beams, Vortex beams and Airy breams. Airy beams exhibit non-diffraction property that can propagate over many Rayleigh lengths without any appreciable change in intensity profiles [29]. Moreover, Airy beams resist diffraction while their main intensity maxima or lobes tend to transversely accelerate during propagation along parabolic trajectories in free space. The self-bending behavior persists over long distances, although the center of gravity of these wave packets remains constant and diffraction eventually takes over. It is also shown that Airy beams have the self-healing properties and can self-reconstruct after propagating through obstacles and retain their intensity profiles under turbulent conditions [30]. Therefore, the non-spreading, self-bending and self-healing properties of Airy beams can make them serve as robust propagation beams resilient against atmospheric turbulence.

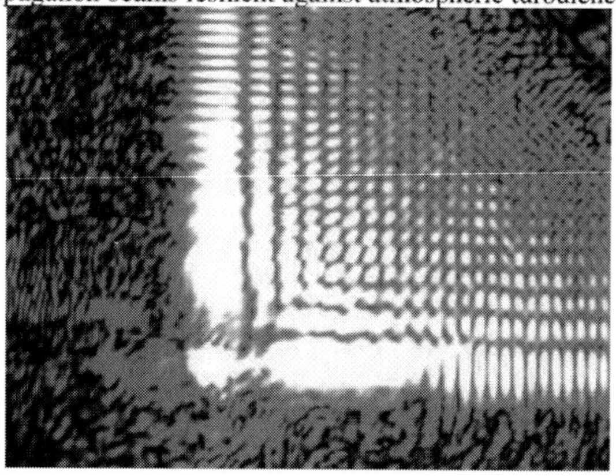

Figure 10. Self-healing of Airy beam propagation through obstructions.

The irradiance and phase distribution of Airy beam are controlled and generated by using a Spatial Light Modulator (SLM) or a hologram phase screen. The SLM is displayed by the Airy beam hologram that consists of the superposition of a 2D phase mask and a diffraction grating. The first order of diffraction is then Fourier transformed using a lens to generate Airy beams. The generated amplitude and phase distribution of Airy beam determine the self-bending and self-healing propagation properties through free space, scattering media and turbulent media as shown in Figure 10. The performance optimization as a function of beam parameters such as coherence length, characteristic length, aperture coefficient and deflection coefficient can be explored to mitigate intensity fluctuation and improve free space optical communication performance.

V. Conclusion and Future Work

The effective integration of high speed, power efficient, compact and reliable free space optical communication terminals with small satellite platforms can pave the way to satellite constellations telecommunication networks. This paper discussed the current architecture limitations and explore the critical technologies, such as compact wide field of view optical antenna, MEMS gimbal-less retroreflective acquisition and tracking mechanism, spatial diversity and beam shaping techniques for atmospheric channel. Further research will focus on long distance MEMS reconfigurable free space optical communication to address environment effects such as LOS blockages and intensity fluctuations. These technologies would help explore the landscape of miniaturized high capacity rugged free space optical communication systems for future satellites networks.

Acknowledgment

We thank professors M. Kavehrad, Z. Liu, H. Gupta, J. Longtin, for sharing their expertise, and financial support by National Science Foundation (NSF) Award # 1513764.

References

[1] O. L. De Weck, R. D. Neufville, and M. Chaize, "Staged Deployment of Communications Satellite Constellations in Low Earth Orbit," *Journal of Aerospace Computing, Information, and Communication*, vol. 1, pp. 119-136, 2004.

[2] K. Woellert, P. Ehrenfreund, A. J. Ricco, and H. Hertzfeld, "Cubesats: Cost-effective science and technology platforms for emerging and developing nations," *Advances in Space Research*, vol. 47, pp. 663-684, Feb. 2011.

[3] V. W. S. Chan, "Free-Space Optical Communications," *Journal of Lightwave Technology*, vol. 24, pp. 4750-4762, 2006.

[4] J. Yin *et al.*, "Satellite-based entanglement distribution over 1200 kilometers," *Science*, vol. 356, pp. 1-5, 2017.

[5] V. W. S. Chan, "Optical satellite networks," *Journal of Lightwave Technology*, vol. 21, pp. 2811-2827, 2003.

[6] P. Deng, "Studies of Novel Partially Coherent Beams propagation through Turbulent Atmosphere," Doctoral Dissertation, Huazhong University of Science & Technology, 2013.

[7] T.-H. Ho, S. D. Milner, and C. C. Davis, "Pointing, acquisition, and tracking system with omnivision," *Free-Space Laser Communications V*, SPIE, 2005, pp. 589219-12.

[8] P. Deng, X. H. Yuan, and D. Huang, "Scintillation of a laser beam propagation through non-Kolmogorov strong turbulence," *Optics Communications*, vol. 285, pp. 880-887, 2012.

[9] P. Deng, X. Yuan, Y. Zeng, M. Zhao, and H. Luo, "Influence of wind speed on free space optical communication performance for Gaussian beam propagation through non Kolmogorov strong turbulence," *Journal of Physics: Conference Series*, 2011, pp. 012056.

[10] P. Deng, X. Yuan, M. Kavehrad, M. Zhao, and Y. Zeng, "Off-axis catadioptric fisheye wide field-of-view optical receiver for free space optical communications," *Optical Engineering*, vol. 51, pp. 063002-11, 2012.

[11] P. Deng, T. Kane, and O. Alharbi, "Reconfigurable free space optical data center network using gimbal-less MEMS retroreflective acquisition and tracking," *Free-Space Laser Communication and Atmospheric Propagation XXX*, SPIE, Feb 2018, pp. 1052403-9.

[12] P. Deng, M. Kavehrad, Z. Liu, Z. Zhou, and X. Yuan, "Capacity of MIMO Free Space Optical communications for multiple partially coherent beams propagation through non-Kolmogorov strong turbulence," *Optics Express*, vol. 21, pp. 15213-15229, Jun 2013.

[13] K. Takahashi and Y. Arimoto, "Compact optical antennas using free-form surface optics for ultrahigh-speed laser communication systems," *Optical Engineering*, vol. 47, pp. 1-9, Jan. 22 2008.

[14] T. Weyrauch, M. A. Vorontsov, G. W. Carhart, G. V. Simonova, L. A. Beresnev, and E. E. Polnau, "Adaptive optical antennas: design and evaluation," *Atmospheric Optics: Models, Measurements, and Target-in-the-Loop Propagation*, SPIE, 2007, pp. 1-12.

[15] G. Druart *et al.*, "Compact infrared pinhole fisheye for wide field applications," *Applied Optics*, vol. 48, pp. 1104-1113, 2009.

[16] P. Deng, X. Yuan, Y. Zeng, and M. Zhao, "Design and evaluation of wide field-of-view optical antenna," *Proc. of SPIE*, vol. 8129, pp. 1-10, 2011.

[17] P. Deng, M. Kavehrad, and X. Yuan, "Comparing wide field-of-view optical receivers for free space optical communications," Photonics Society Summer Topical Meeting Series, IEEE, July 2012, pp. 84-85.

[18] S. M. Navidpour, M. Uysal, and M. Kavehrad, "BER performance of free-space optical transmission with spatial diversity," *IEEE Transactions on Wireless Communications*, vol. 6, pp. 2813-2819, Aug 2007.

[19] P. Deng and M. Kavehrad, "Adaptive Real-Time Software Defined MIMO Visible Light Communications using Spatial Multiplexing and Spatial Diversity," *Wireless for Space and Extreme Environments (WiSEE), 2016 IEEE International Conference on*, 26-29 Sep. 2016.

[20] G. Yun and M. Kavehrad, "Spot-diffusing and fly-eye receivers for indoor infrared wireless communications," *IEEE International Conference on Selected Topics in Wireless Communications*, 25-26 Jun 1992, pp. 262-265.

[21] J. M. Kahn, R. You, P. Djahani, A. G. Weisbin, B. K. Teik, and A. Tang, "Imaging diversity receivers for high-speed infrared wireless communication," *IEEE Communications Magazine*, vol. 36, pp. 88-94, Dec 1998.

[22] P. Deng, M. Kavehrad, and Y. Lou, "MEMS-based beam steerable free space optical communication link for reconfigurable wireless data center," *Broadband Access Communication Technologies Xi*, Proc. SPIE, Feb 2017, pp. 1012805-9.

[23] V. Milanovic, G. A. Matus, and D. T. McCormick, "Gimbal-less monolithic silicon actuators for tip-tilt-piston micromirror applications," *IEEE Journal of Selected Topics in Quantum Electronics*, vol. 10, pp. 462-471, 2004.

[24] M. I. S. Chowdhury, M. Kavehrad, W. Z. Zhang, and P. Deng, "Combined CATV and very high speed data transmission over a 1550-nm wavelength indoor optical wireless link," *Proc. of SPIE*, Feb 2014, pp. 901009-9.

[25] E. Bayaki, R. Schober, and R. K. Mallik, "Performance Analysis of MIMO Free-Space Optical Systems in Gamma-Gamma Fading," *IEEE Transactions on Communications*, vol. 57, pp. 3415-3424, Nov 2009.

[26] J. C. Ricklin and F. M. Davidson, "Atmospheric turbulence effects on a partially coherent Gaussian beam: implications for free-space laser communication," *J. Opt. Soc. Am. A*, vol. 19, pp. 1794-1802, 2002.

[27] A. García-Zambrana, C. Castillo-Vázquez, and B. Castillo-Vázquez, "Outage performance of MIMO FSO links over strong turbulence and misalignment fading channels," *Optics Express*, vol. 19, pp. 13480-13496, 2011.

[28] D. K. Borah and D. G. Voelz, "Spatially partially coherent beam parameter optimization for free space optical communications," *Optics Express*, vol. 18, pp. 20746-20758, 2010.

[29] G. A. Siviloglou, J. Broky, A. Dogariu, and D. N. Christodoulides, "Observation of accelerating airy beams," *Physical Review Letters*, vol. 99, Nov 23 2007.

[30] J. Broky, G. A. Siviloglou, A. Dogariu, and D. N. Christodoulides, "Self-healing properties of optical Airy beams," *Optics Express*, vol. 16, pp. 12880-12891, 2008.